Practical Pharmaceutics

Prof. R.S. GAUD
M.Pharm., Ph.D., F.I.C.
Advisor, AICTE, New Delhi
Formerly, Professor of Pharmacy, SGSITS, Indore,
Dean, Faculty of Technology,
Devi Ahilya Vishwavidyalaya, Indore

Dr. G.D. GUPTA
M.Pharm., Ph.D.
Reader in Pharmaceutics
B.N. College of Pharmacy, Udaipur (Rajasthan)

CBSPD
CBS Publishers & Distributors Pvt Ltd

New Delhi • Bengaluru • Chennai • Kochi • Kolkata • Lucknow • Mumbai
Hyderabad • Jharkhand • Nagpur • Patna • Pune • Uttarakhand

Disclaimer

Science and technology are constantly changing fields. New research and experience broaden the scope of information and knowledge. The authors have tried their best in giving information available to them while preparing the material for this book. Although, all efforts have been made to ensure optimum accuracy of the material, yet it is quite possible some errors might have been left uncorrected. The publisher, the printer and the authors will not be held responsible for any inadvertent errors, omissions or inaccuracies.

PRACTICAL PHARMACEUTICS

ISBN: 978-81-239-0787-1

First Edition: 2002

Reprint: 2004, 2005, 2006, 2007, 2009, 2011, 2016, 2018, 2019, 2020, 2023, 2024

Published by Satish Kumar Jain and produced by Varun Jain for

CBS Publishers & Distributors Pvt Ltd

4819/XI Prahlad Street, 24 Ansari Road, Daryaganj, New Delhi 110 002, India

Ph: 011-23289259, 23266861   Website: www.cbspd.com

e-mail: delhi@cbspd.com

Corporate Office: 204 FIE, Industrial Area, Patparganj, Delhi 110 092

Ph: 011-4934 4934   Fax: 011-4934 4935   e-mail: publishing@cbspd.com

publicity@cbspd.com

Branches

- **Bengaluru:** Seema House 2975, 17th Cross, KR Road, Banasankari 2nd Stage, Bengaluru 560 070, Karnataka, India
  Ph: +91-80-26771678/79   Fax: +91-80-26771680   e-mail: bangalore@cbspd.com
- **Chennai:** 7, Subbaraya Street, Shenoy Nagar, Chennai 600 030, Tamil Nadu, India
  Ph: +91-44-26680620, 26681266   Fax: +91-44-42032115   e-mail: chennai@cbspd.com
- **Kochi:** 42/1325, 1326, Power House Road, Opp KSEB, Power House, Ernakulam 682 018, India
  Ph: +91-484-4059061-65   Fax: +91-484-4059065   e-mail: kochi@cbspd.com
- **Kolkata:** 147, Hind Ceramics Compound, 1st Floor, Nilgunj Road, Belghoria, Kolkata 700 056, West Bengal, India
  Ph: +91-9096713055/56   e-mail: kolkata@cbspd.com
- **Lucknow:** Basement, Khushnuma Complex, 7-Meerabai Marg (behind Jawahar Bhawan), Lucknow 226 001, India
  Ph: +91-522-4000032   e-mail: tiwari.lucknow@cbspd.com
- **Mumbai:** PWD Shed. Gala no. 25/26, Ramchandra Bhatt Marg, Next to JJ Hospital Gate no. 2, Opp. Union Bank of India Noorbaug Mumbai 400 009, Maharashtra, India
  Ph: +91-22-66661880/89   e-mail: mumbai@cbspd.com

Representatives

- **Hyderabad** 0-9885175004 • **Jharkhand** 0-9811541605 • **Nagpur** 0-9421945513
- **Patna** 0-9334159340 • **Pune** 0-9664372571 • **Uttarakhand** 0-9716462459

Printed at Sanjay Printer, Sahibabad, UP, India

Preface

Pharmaceutics is a branch of pharmacy which mainly deals with the principle of dosage form design applicable in the field of pharmaceutical and allied industries due to its major role in developing suitable dosage form. It evolved as a separate subject in due course of time.

Though number of textbooks are available for pharmaceutics, the practical aspect is still found to be area of weakness and one has to depend on various sources. Authors have made an attempt to provide guidelines for practicals to be carried out in the laboratory as per the syllabi of pharmacy education regulating bodies.

The book has been divided into twenty two chapters covering all conventional dosage forms and unit operations. The principle involved, theoretical background, procedure, precautions to be taken in the laboratory and viva voce have been included.

Authors feel that this book will be useful and knowledgeable for diploma and degree students of pharmacy, in understanding basic principles involved in practical pharmaceutics.

Authors are thankful to Mr. H.S. Poplai, General Manager, M/s CBS Publishers & Distributors and his team for their sincere efforts in publishing this book in the present form.

Constructive suggestions, if any, will be appreciated.

Prof. R.S. Gaud
Dr. G.D. Gupta

Contents

1

Aromatic Waters

- Chloroform water
- Concentrated Cinnamon water
- Concentrated Dill water
- Peppermint water
- Concentrated peppermint water
- Cinnamon water
- Camphor water
- Concentrated Anise water
- Chloroform double strength water

Aromatic waters are clear, saturated aqueous solutions of volatile oils or contain other aromatic or volatile substances, also known as medicated waters. Taste and odors of aromatic waters are similar to the drugs or volatile substances from which they are prepared. Aromatic waters should be free from smoky conditions and should be free from unpleasant odors. Generally aromatic waters are used as a flavoring agent, solvent, perfumed vehicles or as a pharmaceutical aid. All substance used in the preparation of aromatic waters should be of pharmacopoeial quality. If the aromatic waters are non-official then it should have best quality of flavoring agents.
Two official methods are commonly used for the preparation of aromatic waters e.g.

1. Solution Method
2. Distillation Method

1. **(a) Solution Method** (without using distributing agent) : Aromatic waters may be prepared by repeatedly shaking 0.2% V/V or 0.2% W/V of volatile substance with purified water. Keep it aside for 12 hours and make up the volume upto 100 ml by adding sufficient purified water after filtration. e.g. chloroform water, camphor water etc.

Advantages

1. Minimum chemicals are required
2. No need of any specific equipment
3. Less time consuming
4. More economic

Disadvantages

1. Light and air may deteriorate volatile oils.
2. Uniform distribution of drug may not be possible.

1. (b) Solution Method (using distributing agents) : Aromatic waters may be prepared by thoroughly shaking the volatile oils with 1.5% of powdered talc or with a sufficient quantity of kieselguhr or pulped filter papers with half quantity of purified water. The mixture is agitated for ten to fifteen minutes by wrist action shaker or by hand. This water is then filtered through the wetted filter papers and volume of the aromatic water is adjusted by the distilled water through the same filter papers.

Advantages

1. Oils or other substances are subdivided by this method
2. Finally divided oils rapidly dissolves and gives saturated solution
3. Repeated shaking of aromatic water is not required
4. Time required is less then other methods
5. Excess oil drops are coalesce which facilitated that filtration of aromatic water
6. It is more economic

Disadvantages

1. It needs more additives
2. It is not suitable to all types of aromatic waters
3. Some times it fails to distribute oils in aqueous part

   e.g. Dill water, Peppermint water, Cinnamon water, Anise water

1. (c) Solution method (with solubilising agents) : Some of the aromatic water are prepared by using solubilising agents such as polysorbate-20 or tween-20 but this method is not convenient because these materials (solubilising agents) are susceptible to moulds growth and in higher concentration (more than 2.0%) impart an objectionable oily taste.

2. Distillation Method : Distillation method for the preparation of aromatic water is the ancient method. Different authorities give different direction for the preparation of aromatic water by the distillation method. Drug distillate ratios for fresh drugs, for dried drugs such as cinnamon, anise Dill, Fennel and Caraway, and for dried leaf such as peppermint are 1:2 or 2:1, 1:10 and 3:10 respectively. Various types of distillation apparatus are used in the distillation process but for the preparation of aromatic waters mainly metallic distillation apparatus is used. Some times steam still is also preferred. Drug material is reduced in small size and added small quantity of purified water. In this process excess of oil substance is removed and, if necessary, the clear water portion is filtered. The distillate is returned several times to the still with fresh portion of the flowers or other material and, used for double distillation, triple distillation quadruple distilled. This process is called cohabation e.g. rose water etc.

Advantages

1. Oil is mixed homogeneously or saturated with water easily.
2. It is suitable for all types of material.
3. Maximum oil extracted from the plants.
4. It is fast process.
5. It is more economic.

Disadvantages

1. Oil gets evaporation by heat.
2. Some unpleasant empyreumatic odor is produced.
3. It is a complicated process.

INCOMPABILITY

Most of the aromatic water containing preparations are incompatible because it is easy to salt-out by the addition of some salts. If aromatic water is used as a vehicle then it may be replaced by distilled water. The salting out action of aromatic water by the action of soluble salts may be used in the evaluation of aromatic waters. This evaluation is based on the amount of standard sodium citrate solution required to produce cloudiness in aromatic water.

PRESERVATION

No preservatives should be added in the aromatic water or medicated waters. If these products are become cloudy or get deteriorated, these products are not suitable to use and should be discarded.

CONTAINERS

1. Use small mouth amber colored glass bottle (50 ml – 500 ml) for packing.
2. Plastic container, screw capped white or colored bottle or may also be used metal containers, small mouth screw capped containers.
3. Containers should be airtight and light resistant.

STORAGE

Product (containers) should be well protected from the light and excessive heat.

LABEL

Label should have following information about the product :

a. Quantity and name of product
b. Composition : Ingredients should be indicated in term of percent or in suitable forms (or each 5 ml contains)
c. Manufacturers name (Mfg. by)
d. Date of manufacturing (Date of Mfg.)
e. Expiry Date (if any)
f. Batch no. and storage condition
g. Mention dose and other special remarks such as warning or cautions etc. where required as mentioned in official books or for maintenance of efficacy of the product e.g.

FOR INTERNAL USE ONLY

FOR EXTERNAL USE ONLY

PROTECT FROM LIGHT

STORE IN DRY AND COOL PLACE

SHAKE WELL BEFORE USE

SHOULD NOT BE USED IF CONTAINS ANY CLOUDY OR FOREIGN PARTICLES

Note : Product should be discarded if they become cloudy or contain any foreign matter by any reason such as deterioration, volatilization, decomposition or mold growth.

EXERCISE NO. 1

Object : To prepare Chloroform water

Requirements : Chloroform, purified water, etc.

Theory : Chloroform (trichloromethane), is clear, odorless and a mobile liquid with ethereal odor and has burning sweet taste. Specific gravity of chloroform is between 1.474 to 1.478. Care should be taken in handling chloroform because in presence of flame or light it produces harmful gases like hydrogen chloride and phosgene gas. The small amount of ethanol in chloroform retards the formation of carbonyl chloride ($CoCl_2$). It is rarely used as an inhalation anesthetic due to its serious side effect on heart and liver. Internally in small dose it is used as carminative and externally as irritant. In the preparation of chloroform water, vigorous shaking is necessary to sub-divide the chloroform in small globules to enhance the solubility. In this preparation distributive agent is not required, as the product is only half saturated with chloroform.

Formula

Chloroform		0.25 ml
Purified water	sufficient to produce	100.0 ml

Procedure

1. Measure the required amount of chloroform and dissolve it in purified water by vigorous shaking.
2. Chloroform is completely dissolved in water.
3. Transfer in clean amber color container and close it tightly.

Category : Pharmaceutical aid.

Dose : 15 to 30 ml

Storage : Store in a well closed container and protected from light.

Uses : Pharmaceutical aid, preservative, vehicle and general anaesthetic.

VIVA VOCE QUESTIONS

1. What is the use of chloroform water?
2. Why chloroform is not used as an anaesthetic agent?
3. How to avoid formation of phosgene gas in chloroform?
4. Why ethanol is used in this preparation?

EXERCISE NO. 2

Object : To prepare Double strength Chloroform water

Requirements : Chloroform, distilled water, etc.

Formula

Chloroform		0.50 ml
Purified water	sufficient to produce	100.0 ml

Procedure

1. Measure required amount of chloroform and dissolve it in purified water by vigorous shaking.
2. Chloroform is completely dissolved in water by repeatedly shaking the solution.
3. Transfer aromatic water in well-cleaned amber color container and close it tightly.

Category : Pharmaceutical aid.

Dose : 5 to 15 ml

Storage : Store in a well closed container and protect from the light.

Uses : Pharmaceutical aid, preservative, vehicle and general anaesthetic.

VIVA VOCE QUESTIONS

1. What is the need of double strength chloroform water?
2. Give application of chloroform water.
3. Why chloroform is not used as an anaesthetic agent?
4. How to reduce formation of phosgene gas in chloroform?
5. Give advantages of double strength chloroform water.
6. What is the difference between chloroform water and double strength chloroform water?

EXERCISE NO. 3

Object : To prepare Camphor water

Requirements : Camphor, ethanol, purified water

Theory : Camphor is a ketone, obtained from *Cinnamomum camphora*. It is colorless or white crystals, granules or crystalline masses. It is highly volatile in nature at room temperature and readily burns with a bright smoky flame. It is soluble in 800 parts of water and in 1 parts of ethanol. In the preparation of aromatic water ethanol acts as distributing agent. Water can not be added in alcoholic solution of camphor because if whole camphor is precipitated then it becomes difficult to re-dissolved it by shaking the solution.

Formula

Camphor	0.10 g
Ethyl alcohol (90%)	0.20 ml
Purified water sufficient to produce	100.0 ml

Procedure

1. Dissolve measured amount of camphor in ethanol.
2. Add alcoholic solution drop wise in purified water with shaking.
3. Afterwards shake occasionally until all the camphor is dissolved
4. If necessary, filter the solution and store in a closed container.

Category : Pharmaceutical aid

Dose : 15 to 30 ml

Storage : Store in a well closed container.

Uses : Pharmaceutical aid, vehicle and used as carminative and to relieve griping pain.

VIVA VOCE QUESTIONS

1. What is camphor? Give pharmaceutical application camphor water?
2. Give advantages of camphor water over the camphor powder.
3. How camphor water is prepared?
4. What is the use of ethanol in water?
5. Why camphor gets precipitation by addition of purified water?
6. Why it is stored in well close container?

EXERCISE NO. 4

Object : To prepare Concentrated Camphor water

Requirements : Camphor, ethanol, and purified water

Formula

Camphor	4.0 g
Ethyl alcohol (90%)	60.0 ml
Purified water sufficient to produce	100.0 ml

Procedure

1. Dissolve measured amount of camphor in ethanol.
2. Alcoholic solution is added drop-wise in purified water with shaking.
3. Afterwards shake occasionally until camphor is completely dissolved.
4. If necessary filter the solution and store in a closed container.

Category : Pharmaceutical aid

Storage : Store in a well closed container.

Uses : Pharmaceutical aid, vehicle and used as carminative and to relieve griping pain.

VIVA VOCE QUESTIONS

1. What is camphor? Mention its importance in pharmacy?
2. Why you prepare concentrated camphor water?
3. Give advantages of camphor water over the camphor powder.
4. What is the difference between concentrated camphor water and water of camphor?
5. What is the use of ethanol in water?
6. Why precipitation is formed if purified water is added?
7. Why it is stored in well-closed container?
8. If camphor is precipitated by addition of water. can it be re-dispersed by shaking.

EXERCISE NO. 5

Object : To prepare Concentrated Cinnamon water

Requirements : Cinnamon oil, ethanol, talc and purified water

Theory : Cinnamon is the dried inner bark of the shoots of coppiced trees of *Cinnamomum zeylanicum*. It contains not less than 1.0 percent of volatile oils. The volatile oil distilled with steam from the leaves and twigs of *Cinnamomum cassia*. It mainly contains yellowish or brownish liquid cinnamaldehyde. It becomes darker and thicker on aging or exposure to the air. It has characteristic odor and taste and specific gravity 1.045 to 1.063.

Formula :

Cinnamon oil	2.0 ml
Ethyl alcohol (90%)	60.0 ml
Purified water sufficient to produce	100.0 ml
Talc powder	5.0 g

Procedure

1. Dissolve measured amount of cinnamon oil in ethanol by shaking.
2. Add small amount of purified water with vigorous shaking.
3. After each addition of water shake it and make volume upto 100.0 ml.
4. Add sterilized purified talc powder and shake it.
5. Allow to stand this solution for a few hours with occasional shaking.
6. Filter the solution and store in a closed container.

Category : Carminative

Dose : 2 to 8 ml

Storage : Preserve in a well-closed container and store in a cool place.

Uses : Carminative and vehicle

VIVA VOCE QUESTIONS

1. What is cinnamon?
2. What is the use of cinnamon water?
3. Differentiate between aromatic water and purified water.
4. Give storage condition of cinnamon water.

EXERCISE NO. 6

Object : To prepare Concentrated Peppermint water IP

Requirements : Peppermint oil, ethanol, talc, and purified water.

Theory : Concentrated aromatic waters are alcoholic saturated solutions of volatile oils or other substances and about 40 times stronger than the aromatic waters. Aromatic waters are prepared by diluting 1 part of concentrated aromatic waters with 39 parts of the purified water. In these preparations oil content is excess than that required to form a saturated solution. Purified talc or French chalk is used as an absorbent to remove excess amount of oil. When the preparation is allowed to stand for a few hours to finely divided globules of oil to coalesce and shake it to enhance the absorption of the excessive or insoluble amount of oil. The aromatic part of the oil is soluble in the vehicle while other non-aromatic part containing terpenes is insoluble in the vehicle. The undissolved non-aromatic part of the oil is separated along with the talc by filtration of water.

Formula

Peppermint oil	2.0 ml
Ethyl alcohol (90%)	60.0 ml
Purified water sufficient to produce	100.0 ml

Procedure

1. Dissolve measured amount of volatile oil in 60.0 ml of 90 percent ethanol.
2. Add sufficient purified water in successive small portion with vigorous shaking and make volume 100.0 ml.
3. It is necessary to shake vigorously after each addition
4. Add 5.0 g of sterilized purified talc and shake it.
5. Allow to stand for a few minutes and shake occasionally.
6. Remove talc by filtration and store in a well-closed container.

Category : Flavoring agent **Dose :** 0.25 to 1.0 ml

Storage : Store in a well closed container

Uses : It is used as flavoring agent and carminative.

VIVA VOCE QUESTIONS

1. What is peppermint? Give application of talc in peppermint water?
2. How to prepare peppermint water?
3. Give pharmaceutical application of aromatic water.
4. What is the use of ethanol in water?

EXERCISE NO. 7

Object : To prepare Concentrated Dill water

Requirements : Dill oil, ethyl alcohol, talc, and distilled water

Theory : Dill (sowa) consists of the dried fruits of *Anethum graveolens* (Fam. Umbelliferae). It has characteristic aromatic odor and taste. It contains about 2.5 percent of volatile oil. When dill oil is dissolved in ethanol and purified water is added, the non-aromatic terpene of dill oil gets precipitated. For absorption of precipitated terpenes, talc powder is added in the solution and filtered to get a clear and bright solution.

Formula

Dill oil	2.0 ml
Ethyl alcohol (90%)	60.0 ml
Purified water sufficient to produce	100.0 ml
Talc powder	5.0 g

Procedure

1. Dissolve measured amount of dill oil in ethanol.
2. Add purified water in successive small quantity with continuous shaking.
3. Make up the volume 1000 ml by purified water with continuous shaking.
4. Add 50 g of talc powder shake it and keep aside for few hours.
5. Shake the solution and filter it.
6. Clear and bright solution obtained is stored in a closed container.

Category : Carminative **Dose :** 0.3 to 1.0 ml

Storage : Preserve dill water in a well closed container, protected from light and store in a cool place (temperature 25°C)

Uses : Carminative and flavoring agent

VIVA VOCE QUESTIONS

1. What is dill oil?
2. How to prepare dill water?
3. What is the role of talc in dill water?
4. Why dill water is to be stored in a cool place?
5. Give advantages of dill water?
6. Give pharmaceutical application of concentrated aromatic dill water.
7. How to store dill water? Give ideal condition of storage.

EXERCISE NO. 8

Object : To prepare Concentrated Anise water

Requirements : Anise oil, ethanol, talc and purified water

Formula

Anise oil	2.0 ml
Ethyl alcohol (90%)	60.0 ml
Purified water sufficient to produce	100.0 ml
Talc powder	5.0 g

Procedure

1. Dissolve measured amount of anise oil in ethanol.
2. Add purified water in successive small quantity with continuous shaking.
3. Make up the volume 100.0 ml by purified water with continuous shaking.
4. Add 5.0 g of talc powder shake it and keep aside for few hours.
5. Shake the solution and filter it.
6. Clear and bright solution obtained is stored in a closed container in dark place.

Category : Pharmaceutical aid

Dose : 2 to 8 ml

Weight per ml : 0.978 to 0.988 at 25°C

Storage : Preserve in a well-closed container, protected from light and store in a cool place.

Uses : Flavoring agent and carminative

VIVA VOCE QUESTIONS

1. What is anise oil?
2. How to prepare anise water?
3. What is the role of talc in preparing anise water?
4. Which material is used as a absorbent to remove excess amount of oil?
5. Give pharmaceutical application of concentrated aromatic anise water.

EXERCISE NO. 9

Object : To prepare Concentrated Caraway water

Requirements : Caraway oil, ethanol, talc, and purified water

Formula

Caraway oil	2.0 ml
Ethyl alcohol (90%)	60.0 ml
Purified water sufficient to produce	100.0 ml
Talc powder	5.0 g

Procedure

1. Dissolve measured amount of caraway oil in ethanol.
2. Add purified water in successive small quantity with continuous shaking.
3. Make up the volume 100.0 ml by purified water with continuous shaking.
4. Add 5.0 g of talc powder, shake it and keep aside for few hours.
5. Shake the solution and filter it.
6. Clear and bright solution store in a closed container at dark place.

Category : Carminative

Dose : 2 to 8 ml

Weight per ml : 0.900 to 0.917 g at 25°C

Storage : Preserve in a well-closed container, protected from air and light and store in a cool place.

Uses : Carminative and flavoring agent

VIVA VOCE QUESTIONS

1. What is the source of caraway?
2. How to prepare caraway water?
3. What is the role of talc in caraway water?
4. Which material is used as an absorbent to remove excess amount of oil?
5. Give pharmaceutical application of concentrated aromatic anise water.

2

Capsules

- Ibuprofen capsule
- Amoxycillin capsule
- Chlortetracycline capsule
- Paracetamol capsule
- Diphenhydramine capsules

In 1834, Mothes and Dublanc invented single-piece, olive-shaped, gelatin capsules which were closed after filling using a drop of concentrated warm gelatin solution. In 1848, James Murdock invented two piece telescoping capsule. The word capsule is derived from the Latin 'capsula' meaning a small box. The word is used to describe an edible package made from gelatin that is filled with medicines to produce a unit dose mainly for oral use. In another word, 'capsules are solid dosage form in which one or more medicament with or without inert substances are enclosed within a small gelatin shell

The capsule shells are composed of gelatin or other materials, the consistency of which may be adjusted by the addition of substances such as glycerol and sorbitol or a mixture of both. It contains excipients such as surface-active agents, opaque fillers, anti-microbial preservatives, sweetening agents, flavoring agents and coloring agents. The contents of capsules may be of solid, liquid or paste-like consistency. Capsules consist of one or more medicaments, with or without excipients such as solvents, diluents, lubricants or disintegration of the shell.

There are two types of capsules

I. HARD GELATIN CAPSULES

It consists of mainly two pieces, a cap and a body that fit inside the other. It is manufactured from a mixture of gelatin, colorants and sometimes may contain opacifying agents and is filled in separate operation. Empty capsules are sold by sizes from 000 the largest to size 5, the smallest. The most popular sizes in practice are size 0 to 5 for human use.

Different Size of capsules and volume filled in ml

S.No.	Capsule No.	Volume in ml
1.	000 Largest size	1.36
2.	00	0.95
3.	0	0.67
4.	1	0.48
5.	2	0.37
6.	3	0.27
7.	4	0.20
8.	5 Smallest size	0.13

Capsule shells are prepared by two methods such as:

1. Plate Process 2. Rotary Die Process

Following material are filled in hard gelatin capsules :

1. Dry solids : Powder, granules, pellets, tablets
2. Semisolid : Thixotropic mixture, thermosoftening mixtures, pastes, etc.

Advantages of Capsules

1. Ease of use due to the fact that it is smooth, slippery and easy to swallow and tasteless for drugs.
2. Suitable for substances having bitter taste and unpleasant odor.
3. As produced in large quantities it is economic and attractive as available in wide range of colors.
4. Minimum excipients required.
5. Little pressure required to compact the material.
6. Unit dosage form.
7. Easy to store, and transport.

Disadvantages of Capsules

1. Not suitable for highly soluble substances like potassium chloride, potassium bromide, ammonium chloride, etc.
2. Not suitable for highly efflorescent or deliquescent materials.
3. Special conditions are required for storage.

Determination of Capsule Fill Weight

Capsule fill weight is determined applying following formula

Capsule filling weight = Tapped bulk density of substance or powder × capsule volume

Example- I

A formulation has a theoretical fill weight of 325 mg and tapped bulk density of substance 0.75 g/ml while bulk density of inert substance is 0.80 g/ml.

Given

1. Weight of substance in each capsule = 325.0 mg
2. Bulk density of substance = 0.75 g/ml
3. Bulk density of inert substance = 0.80 g/ml

$$\text{Volume occupied by fill weight} = \frac{\text{Weight of substance in each capsule}}{\text{Bulk density of substance}}$$

$$= \frac{0.325}{0.75} = 0.43 \text{ ml}$$

Volume of size '1' capsule = 0.48 ml (by table 1.)
Volume occupied by drug = 0.43 ml (by calculation)
Volume unoccupied = 0.48 – 0.43 = 0.05 ml

Weight of diluent or inert substances = volume × bulk density
= 0.05 × 0.80 = 0.040 g = 40.0 mg

Send ten capsules of the given materials

Weight for 11 capsule of substance = 325 × 11 = 3.575 g
Weight of diluent or inert substance = 40 × 11 = 0.440 g

Total weight of filling substance in the hard empty shell = 4.015 g

Example II

Fill the substance containing drug in each capsule, of size '1' capsule 280 mg of bulk density 0.70 g/ml. Determine how much amount of inert substance is required to fill 10 capsule while bulk density of inert substance is 0.85.

Solution : similar problem as in example I.

Example III

Which number of capsule is to be used 375 mg material having bulk density 0.75 g/ml. Bulk density of inert substance is 0.68. Calculate the total g/ml required for 10 capsules.

Given

1. Weight of substance in each capsule = 375.0 mg
2. Bulk density of substance = 0.75 g/ml
3. Bulk density of inert substance = 0.68 g/ml

$$\text{Volume occupied by fill weight} = \frac{\text{Weight of substance in each capsule}}{\text{Bulk density of substance}}$$

$$= \frac{0.375}{0.75} = 0.466 \text{ ml}$$

For filling this volume of substance '1' number capsule is most appropriate because

Volume of size '1' capsule = 0.48 ml (by table 1.)
Volume occupied by drug = 0.466 ml (by calculation)
Volume unoccupied = 0.48 – 0.466 = 0.014 ml

Weight of diluent or inert substances = volume × bulk density
= 0.014 × 0.68 = 0.040 g = 0.00952 g = 9.52 mg

Send ten capsules of the given materials

Weight for 11 capsule of substance = 375 × 11 = 4.125 g
Weight of diluent or inert substance = 9.52 × 11 = 0.105 g

Total weight of filling substance in the hard empty shell = 4.230 g

---

Note : Always weigh substance for one extra capsule as material is lost during powdering, mixing, and handling of the substance.

Evaluation of Capsule

1. Drug content : Determine by assay process.
2. Weight variations : Determine by simple weighing of 20 tablets using digital balance, should be within the limit of 90% to 110% in weight.
3. Disintegration : Disintegration test is usually not required for capsules unless it is treated to resist solution in gastric fluid. In such cases it must meet the requirements for disintegration of enteric-coated tablets.
4. Dissolution test : Use dissolution apparatus basket assembly for determination of dissolution test.

Storage

Capsules should be kept in well-closed containers not exceeding 30°C.

Filing Equipments

1. Eli Lilly Co., Indianapolis, IN.
2. Farmatic SNC, Bologna, Itlay
3. Hoflinger and Karg, Waiblingen, Germany
4. mG_2 S.P.A., Bologna, Itlay
5. Osaka, Osaka, Japan
6. Macofar SAS, Bologna, Itlay
7. Perry Industries, Green Bay, WI
8. Zanasi Nigris, S.P.A., Bologna, Itlay
9. Parke-davis and Company, Detroit, MI

Other Machines

1. Rotoweigh : Capsule weighing machine
2. Rotosort : Filled capsule sorting machine
3. Seidenader Model PM 60 : Unfilled capsules cleansing and polishing machine
4. Erweka KEA : Capsule deducting and polishing machine
5. Hartnett Model B : Capsule imprinting machine
6. Markem Model 280 A : Capsule imprinting machine

II. SOFT GELATIN CAPSULES

The shells of soft capsules are thicker than those of hard capsules. It consist of a single part and may be of various shapes. The shells are usually formed, filled and sealed in a combined operation but in some cases, shells for extemporaneous use may be performed. The shell material may contain medicament. It may be used for enclosing liquid material and solids are usuallv dissolved or dispersed in a suitable excipient to give a paste like consistency.

EXERCISE NO. 1

Object : To prepare Amoxycillin capsule

Requirements : Amoxycillin trihydrate, diluent, etc

Theory : Amoxycillin is a semi-synthetic penicillin, an analogue of ampicillin, with a broad spectrum of bactericidal activity against many gram-positive and gram-negative micro-organisms. It is stable in presence of gastric acid and may be given without regard to meals. It acts through inhibition of biosynthesis of cell wall mucopeptide.

Formula

Each capsule contain

Amoxycillin trihydrate	250 mg
Diluent	q.s.

Procedure

1. Calculate the quantity of drug substances and diluent as given in example I using the formula and calculate the quantity for one extra capsule.
2. Select the capsule size as per requirement.
3. Fill the capsule carefully by hand operating machine or manually with minimum weight variations.

Category : Antibiotic

Dose : 2 to 3 capsule daily

Storage : Preserve in a well-closed container. The quantity of the active ingredient is stated in terms of the equivalent amount of amoxycillin. The label on the container states.

1. Amount of amoxycillin.
2. Expiry date after which the capsules are not intended to be used.
3. Condition under which they should be stored.

Uses : It is used in respiratory, genito-urinary, skin and soft tissue, ENT infections due to susceptible strains of Gram negative organisms like H. influenzae, E.coli, and N. gonorrhea.

EXERCISE NO. 2

Object : To prepare Ibuprofen capsule

Requirements : Ibuprofen, diluent, etc.

Formula : Each capsule contain

Ibuprofen	300 mg
Diluent	q.s.

Procedure

1. Calculate the quantity of drug substances and diluent as given in example I using the formula and calculate the quantity for one extra capsule.
2. Select the capsule size as per requirements.
3. Fill the capsule carefully by hand operating machine or manually with minimum weight variations.

Category : Analgesic, antipyretic, rheumatoid arthritis

Dose : Ibuprofen 200-600 mg twice daily

Storage : Preserve in a well-closed container.

Uses : Analgesic, antipyretic, rheumatoid arthritis.

EXERCISE NO. 3

Object : To prepare Chlortetracycline capsule

Requirements : Chlortetracycline, diluent, etc.

Formula

Chlortetracycline	250 mg
Diluent	q.s.

Procedure : Similarly as Ex. No. 2

Category : Antibiotic

Dose : 1 to 3 tablets daily in divided doses.

Storage : Preserve Chlortetracycline capsule in a well-closed container in a cool place.

Uses : Antiprotozoal preparation

EXERCISE NO. 4

Object : To prepare Diphenhydramine capsules

Requirements : Diphenhydramine, diluent, etc.

Formula

Diphenhydramine	50 mg
Diluent	q.s.

Procedure : Similarly as in Ex. No. 2

Category : Antihistamine

Dose : 1 to 2 tablets daily in divided dose

Storage : Preserve in diphenhydramine capsules in a cool place.

Uses : Antihistamine

EXERCISE NO. 5

Object : To prepare Paracetamol capsule

Requirements : Paracetamol, diluent, etc.

Formula

Paracetamol	250 mg
Diluent	q.s.

Procedure : Similarly as in Ex. No. 2

Category : Analgesic and anti-inflammatory

Dose : 1 to 3 tablets

Storage : Preserve in a well-closed container in cool place

Uses : Analgesic and anti-inflammatory

VIVA VOCE QUESTIONS

For Exercise No 1 to 5

1. What are capsules?
2. How to prepare soft gelatin capsules?
3. Classify hard gelatin capsules.
4. How to calculate quantity of drug in capsule?
5. How to calculate displacement value of drug?
6. What is the difference between hard and soft gelatin capsules?
7. How to fill hard gelatin capsules?
8. Which types of material are suitable to fill in hard gelatin capsules?
9. Define eutectic, hygroscope and deliquescent substances with examples.
10. Which types of materials are not suitable to fill in the capsules?
11. What are the advantages of capsules?
12. Which types of machines are used for the manufacturing of capsules?
13. How to seal the capsules?
14. How to preserve the capsules?
15. What percentage of moisture is essential in hard gelatin capsules?
16. What is the use of glycerin in capsule?
17. How to prepare soft gelatin capsules?
18. Why capsules dosage form is required?
19. Give additives for the preparation of hard gelatin capsules.
20. What is microencapsulation?

3

Creams

- Aqueous cream
- Buffered cream
- Aqueous calamine cream
- Zinc cream
- Salicyclic acid and sulphur cream
- Zinc and Ichthammol cream
- Cold cream
- Vaneshing cream
- Cetrimide cream

The term 'cream' in pharmacy and medicine is applied to viscous emulsions or semi-solid preparations consisting of solutions or dispersions of one or more medicaments in suitable base and intended for application to the skin or mucous membrane. They are applied to the skin for protective, beautifying, therapeutic or prophylactic purposes. Creams may contain suitable antimicrobial or preservatives unless the medicaments or basis have sufficient intrinsic bactericidal and fungicidal activity.

Creams are mainly two types

1. water-in-oil (w/o)
2. oil-in-water (o/w)

DILUTION OF CREAMS

Sometimes dilutions of proprietary creams are required. Care should be taken in diluting creams particularly to prevent microbial contamination. For dilution of creams, appropriate diluent should be used and heating should be avoided during mixing of diluent in the creams. Sometimes use of diluent may affect the stability of creams. Diluted creams should be used within two weeks of their preparation.

The stability and activity of cream may be impaired by the dilution such as :

1. Diluent without antimicrobial agent reduces the overall concentration of the preservative in the preparation.
2. Sometimes creams change o/w to w/o by the diluent property.
3. Change of pH by dilution, may cause precipitation or degradation of medicaments.
4. Choice of wrong diluent may cause poor solubility of medicament.
5. The activity of medicament may be reduced due to complex formation with the diluent.
6. It may inverts the activity of emulsifying agents.

CONTAINER

Cream should be supplied in suitable containers fitted with a closure which reduces contamination of micro-organisms and evaporation of water while the product is not in use. For commercial purpose creams should be supplied in suitable collapsible tubes.

LABELLING

Comply with the general requirements for labeling and should also specify.

1. Expiry date.
2. Storage conditions

STORAGE

Creams should be stored at temperature not exceeding 25°C. They should not be allowed to freeze.

EXERCISE NO. 1

Object : To prepare Aqueous cream

Requirements : Emulsifying wax, white soft paraffin, liquid paraffin, chlorocresol, purified water, etc.

Formula

A.

Emulsifying ointment	30.0 g
Chlorocresol	0.1 g
Purified water, freshly boiled and cooled	69.9 g

B. Emulsifying cream

Emulsifying wax	30.0 g
White soft paraffin	50.0 g
Liquid paraffin	20.0 g

Procedure :

1. Emulsifying ointment is prepared by melting all ingredients together and stir until cold.
2. Dissolve chlorocresol in small amount of purified water with the aid of gentle heat.
3. Melt the emulsifying ointment and add the solution of chlorocresol while still warm.
4. Stir gently until cold.

Category : Pharmaceutical aid

Storage : Preserve in well-closed container (collapsible tube), in cool place.

Uses : Pharmaceutical aid

VIVA VOCE QUESTIONS

1. What is cream?
2. How prepare cream is prepared?
3. What is the use of chlorocresol?
4. How emulsifying ointment is prepared?
5. What is the use of emulsifying cream

Note : If another antimicrobial preservative replaces chlorocresol in this formulation, the suitability of the creams as diluent should be confirmed before use.

EXERCISE NO. 2

Object : To prepare Buffered cream

Requirements : Emulsifying wax, white soft paraffin, liquid paraffin, sodium phosphate, citric acid monohydrate, chlorocresol, purified water, etc.

Formula

Emulsifying ointment	30.0 g
Sodium phosphate	2.5 g
Citric acid monohydrate	0.5 g
Chlorocresol	0.1 g
Purified water, freshly boiled and cooled	66.9 g

Procedure

1 Dissolve sodium phosphate, citric acid monohydrate and chlorocresol in purified water with gently heat.
2 Melt the emulsifying ointment with heat.
3 Heat both solution and emulsifying ointment at the same temperature and mix with stirrer gently until cold.

Category : Pharmaceutical aid

Acidity : pH- 5.7 to 6.3

Storage : Preserve in aluminium tubes, their inner surfaces should be coated with suitable lacquer, at cool place.

Uses : Pharmaceutical aid

VIVA VOCE QUESTIONS

1. What is difference between ointment and paste?
2. How will you prepare cream?
3. Define cream with examples.
4. What is cold cream?
5. What is the relation between cream and emulsion?

EXERCISE NO. 3

Object : To prepare Aqueous calamine cream

Requirements : Calamine, zinc oxide, arachis oil,, emulsifying wax, purified water, etc.

Theory : Calamine was originally obtained by roasting a native zinc carbonate. Calamine contains after ignition, not less than 98.0 percent of zinc oxide. It is pink colour all of which passes through a no. 100 standard mesh sieve. It is odorless and almost testless. It is used as astringent, protective and soothing ointments and lotions for sunburn, ivy poisonong etc.

Formula

Calamine	4.0 g
Zinc oxide	3.0 g
Arachis oil	30.0 g
Emulsifying wax	6.0 g
Purified water, freshly boiled and cooled	57.0 g

Procedure

1. By gentle heat dissolve emulsifying wax in arachis oil.
2. Add 45.0 ml of purified water at the same temperature.
3. Stir until cool.
4. Triturate calamine and zinc oxide with remaining quantity of purified water.
5. Add this mixture in the cream and mix uniformely.

Category : Astringent and protective

Storage : Preserve in a well-closed container, in cool place.

Uses : Astringent and protective

VIVA VOCE QUESTIONS

1. How will you prepare aqueous calamine cream?
2. What is calamine?
3. Why calamine and zinc oxide is used in this preparation?

EXERCISE NO. 4

Object : To prepare Cetrimide cream

Requirements : cetrimide, cetostearyl alcohol, liquid paraffin, purified water, etc.

Theory : Cetrimide consists of a mixture of dodecyl-, tetradecyl- and hexadecyl-trimethylammonium bromides. It is a creamy white, voluminous, free flowing powder, faint odor and has bitter test. It is freely soluble in water but more soluble in warm water and practically insoluble insolvent ether. It is used as a bactericide and astringent.

Formula

Cetrimide	0.5 g
Cetostearyl alcohol	5.0 g
Liquid paraffin	50.0 g
Purified water, freshly boiled and cooled	100.0 g

Procedure

1. Dissolve the cetostearyl alcohol in the liquid paraffin with the aid of gentle heat (A).
2. At the same temperature in another beaker dissolve cetrimide in purified water (B).
3. Add warm oily phase (A) into aqueous phase (B) with stir until cold.

Category : Pharmaceutical aid and bactericide

Storage : Preserve in a well-closed container.

Uses : Pharmaceutical aid and bactericide

VIVA VOCE QUESTIONS

1. What is cetrimide cream?
2. What is the difference between cream and paste?
3. What is emulsion?
4. What is ointment?
5. What is bactericide and anti-bacterial?
6. How to prepare cream?

EXERCISE NO. 5

Object : To prepare Zinc cream

Requirements : Zinc oxide, calcium hydroxide, oleic acid, arachis oil, wool fat etc.

Theory : Zinc oxide is a very fine odorless, amorphous, white or yellowish white powder and free from gritty particles. It gradually absorbs carbon dioxide from air. When it is heated strongly, a yellow color appeares, which disappear on cooling. Its suspension in water is neutral. It reacts with fatty acids in oils and fats to produce lumpy masses of zinc oleate, stearate, etc. In the preparation cream, zinc oxide is levigate to a smooth paste with a little mineral oil before incorporation into the cream or ointment.

It has a mild antiseptic action, astringent, and protective. It is widely employed in the treatment of skin diseases and infections as eczema, impetigo, psoriases, and varicose ulcers.

Formula

Zinc oxide	32.0 g
Calcium hydroxide	0.04 g
Oleic acid	0.5 ml
Arachis oil	32.0 ml
Wool fat	8.0 g
Purified water, freshly boiled and cooled, q.s. to	100.0 g

Procedure

1. Mix weighed quantity of zinc oxide and calcium hydroxide.
2. Mix arachis oil and oleic acid and triturate with the mixture of calcium hydroxide and zinc oxide.
3. Incorporate the wool fat slowly by mixing.
4. Add purified water gradually with continuous stirring.
5. Add sufficient purified water to produce 100.0 g.

Category : Astringent and protective

Storage : Preserve in a well-closed container.

Uses : Astringent and protective

VIVA VOCE QUESTIONS

1. What is the use of zinc?
2. What is astringent?
3. How to prepare and evaluate creams?
4. What is face cream?
5. Cream is a cosmetic or medicated preparation, Explain with suitable example.

EXERCISE NO. 6

Object : To prepare Salicylic acid and sulphur cream

Requirements : Aqueous cream, salicylic acid, precipitated sulphur, etc.

Formula

Salicylic acid, finely sifted	2.0 g
Precipitated sulphur	2.0 g
Aqueous cream, sufficient to produce	100.0 g

Procedure

1. Triturate the salicylic acid and the precipitated sulphur with part of the aqueous cream until smooth cream is formed.
2. Add remainder of the aqueous cream and mix properly.

Category : Protective

Storage : Preserve in a well closed container

Uses : Protective

VIVA VOCE QUESTIONS

1. What is salicylic acid?
2. What is the use of salicylic acid?
3. How salicylic acid cream is prepared?
4. How to use salicylic acid cream?
5. What is the storage condition of cream?

EXERCISE NO. 7

Object : To prepare Zinc and Ichthammol cream

Requirements : Ichthammol, cetostearyl alcohol, wool fat, zinc cream, etc.

Formula

Ichthammol	5.0 g
Cetostearyl alcohol	3.0 g
Wool fat	10.0 g
Zinc cream, sufficient to produce	100.0 g

Procedure

1. Melt weighed amount of the wool fat and cetostearyl alcohol with the aid of gentle heat.
2. Triturate the mixture with zinc cream until it becomes smooth.
3. Incorporate the Ichthammol in the smooth cream.
4. Add sufficient zinc cream to produce 100 g and mix.

Category : Protective

Storage : Preserve in well-closed container.

Uses : Protective

VIVA VOCE QUESTIONS

1. What is Ichthammol?
2. What is the use of Ichthammol?
3. How will you prepare cream of Ichthammol?

EXERCISE NO. 8

Object : To prepare cold cream

Synonyms : Petrolium rose water ointment USP

Requirements : Cetyl esters wax, white wax, mineral oil, sodium borate, etc.

Theory : Cold cream is a water-in-oil emulsion. It is useful as an emollient, cleansing cream, and ointment base. It resembles 'Rose water ointment, differing only in that mineral oil is used in place of almond oil and omitting the fragrance. This change produces an ointment base which is not subjected to rancidity like one containing a vegetable oil.

Formula

Cetyl esters wax	12.5 g
White wax	12.0 g
Mineral oil	56.0 g
Sodium borate	0.5 g
Purified water	19.0 g

Procedure

1. Reduce size of cetyl esters wax and white wax in small pieces.
2. Melt them on a steam bath with the mineral oil and continue heating until the temperature of the mixture reaches 70°C (A).
3. Dissolve the sodium borate in purified water and heat at 70°C in separate flask (B).
4. Gradually add aqueous solution (B) to the mixture of oily phase (A) with continuous and rapid stirring until it has congealed.

Category : Emollient and cleansing cream.

Storage : Preserve in a well-closed container.

Uses : Emollient and cleansing cream.

VIVA VOCE QUESTIONS

1. What is cold cream?
2. What is the mechanism of cold cream?
3. Why cold cream is used in winter? What happened if it is used in the summer?
4. How will you prepare cold cream? Give application of cold cream.

EXERCISE NO. 9

Object : To prepare Vanishing cream

Requirements : Stearic acid, stearyl alcohol, cetyl alcohol, methylparaben, propylparaben, potassium hydroxide, purified water, etc.

Formula

Stearic acid	13.0 g
Stearyl alcohol	1.0 g
Cetyl alcohol	1.0 g
Methylparaben	0.10 g
Propylparaben	0.05 g
Potassium hydroxide	0.90 g
Purified water, sufficient to produce	100.0 g

Procedure

1. Melt stearic acid, stearyl alcohol, and cetyl alcohol at 75°C in a conical flask. (A)
2. Dissolve potassium hydroxide in purified water at 75°C in another conical flask and add preservative. (B)
3. Add (A) into (B) at 75°C slowly with continuous stirring.
4. Slowly stir the mixture until a smooth cream is formed at room temperature.

Category : Protective

Storage : Preserve in a well-closed container.

VIVA VOCE QUESTIONS

1. What is vanisning cream?
2. What is the use of vanishing cream?
3. Vanishing cream is used in summer why?
4. Give storage condition of vanishing cream.
5. Can you categories boroline as a vanishing cream?

4

Ear Drops

- Aluminium acetate ear drops
- Hydrogen peroxide ear drops
- Framycetin ear drops
- Sodium bicarbonate ear drops
- Boric acid ear drops
- Chloramphenicol ear drops
- Ichthammol ear drops
- Phenol ear drops
- Glyceric magnesium sulphate ear drops
- Hydrocortisone and Neomycin ear drops
- Phenol and Amethocaine ear drops
- Sulphacetamide ear drops

Ear drops are liquid preparation and usually are suspension or emulsion or solution of one or more medicaments in vehicles suitable for instillation into the ear. Eye drops are generally safe for the ears even if there is a hole in the eardrum. Eye drops are generally gentle enough for the middle ear. The ingredients of ear drops are different depending on the purpose, 15ml to be dispensed unless otherwise directed.

Ear drops range from liquids designed to soften ear wax, to medicines made to cure bacterial infections, to medicines against fungus. Many kinds of eye drops can be used in the ear but the opposite is *not* true. Ear drops are too harsh for the eyes.

Ear drops fall into two major practical categories

1. Those that are gentle enough to be safe in the part of ear behind the ear drum,
2. Those that are not safe for ear drum.

How to Use Ear Drops

1. Warm the ear drop bottle by holding it in your hands for a few minutes. Shake the bottle well.
2. Unscrew the cap of the bottle and draw some liquid into the dropper.

3. Either lie on your side or tilt your head over so that the ear which needs the drops is facing upwards.
4. Gently pull your earlobe upwards, away from your neck, and squeeze the correct number of drops into the ear.
5. Keep your head tilted for about five minutes so that the drops can spread into the ear.
6. Straighten your head and wipe away any extra liquid with a clean tissue.
7. Replace the cap on the bottle.
8. Store your ear drops in a cool, dark place.
9. You must use your ear-drops for the full length of the treatment course, even if your ear feels better. If you stop too soon, your ear problems may return.
10. While using your ear drops, try not to get water in your ear. Be careful when washing your hair and do not go for swimming until the course is finished.
11. When you have finished your course of treatment, throw any leftover drops away, or return them to your pharmacist for disposal.

Don't mix-up! Eyes drops or ear drops?

A patient put ear drops in eye and experienced burning and itching. Pharmacy dispensed ear drops instead and labeled the drops as eye drops. Unfortunately this is not too uncommon, due to variety of reasons. Sometimes the physician's handwriting is so poor that the pharmacist misreads the prescription. As a patient, be aware that this may happen. Double-check the label on the product if it says "otic," it's for the ear, and if it says "ophthalmic," it's for the eye. Ear drops never go in the eye, but some eye products can be used safely in the ear.

CONTAINERS

Ear drops should be supplied in containers of glass or suitable plastic which are fitted with an integral dropper or with a cap of suitable materials incorporating a suitable dropper tube and rubber or plastic teat. Alternatively, such cap assembly is supplied separately.

LABELLING

Comply with the general requirements for labeling on the ear drops container. It should clearly state that the prescription is intended for external use only.

EXERCISE NO. 1

Object : To prepare Aluminium acetate ear drops

Synonyms : Burow's Solution, Aluminium acetate solution

Requirements : Aluminium sulphate, acetate acid, tartaric acid, calcium carbonate, etc.

Formula

Aluminium sulphate	4.5 g
Acetate acid	5.0 ml
Tartaric acid	0.9 g
Calcium carbonate	2.0 g
Purified water	15.0 ml

Procedure

1. Weigh required amount of aluminium sulphate and dissolve in 10.0 ml of purified water.
2. Add acetic acid and then calcium carbonate.
3. Add remaining amount of purified water and mix properly.
4. Allow to stand for not less than twenty-four hours in a cool place and stirrer occasionally
5. Filter the solution after twenty-four hours.
6. Add tartaric acid in the filtrate and mix.

Category : Astringent and as a local antiperspirant **Dose :** 3 to 5 drops

Storage : Aluminium acetate ear drops should be kept in a well-filled container and stored at a temperature not more than 25°C.

Uses : It is used as a powerful astringent and as a local antiperspirant.

VIVA VOCE QUESTIONS

1. Why ear drops is required?
2. What is the difference between ear drops and eye drops
3. How will you prepare eye drops?
4. It is a sterile preparation or not explains with reasons?
5. Mention various types of ear diseases.
6. Give store condition of ear drops.

EXERCISE NO. 2

Object : To prepare Sodium bicarbonate ear drops

EXERCISE NO. Sodium bicarbonate, glycerin, purified water, etc.

Formula

Sodium bicarbonate	0.75 g
Glycerin	4.5 ml
Purified water, sufficient to produce	15.0 ml

Procedure

1. Dissolve sodium bicarbonate in freshly boiled and cooled purified water in about 6 ml.
2. Add glycerin mix.
3. Add sufficient purified water to produce 10.0 ml.

Weight per ml : 1.10 to 1.12 g

Dose : 3 to 5 drops

Storage : Preserve in a well-filled container and store at a temperature not more than 25°C.

Uses : Antipruritic

VIVA VOCE QUESTIONS

1. What is the use of sodium bicarbonate?
2. How will you prepare ear drop?
3. What is the use of glycerin in ear drops?
4. Define antipruritic.

EXERCISE NO. 3

Object : To prepare Hydrogen peroxide ear drop

Requirements : Hydrogen peroxide, purified water etc.

Formula

Hydrogen peroxide solution (6.0%)	3.75 ml
Purified water, sufficient to produce	15.0 ml

Procedure

1. Mix measured amount of hydrogen peroxide solution with purified water.
2. If any turbidity or impurity persist filter the solution.

Category : Pharmaceutical aid.

Dose : 3 to 5 drops

Storage : Preserve in a well-filled container and stored at a temperature not more than 25° C.

Uses : Soften ear wax

VIVA VOCE QUESTIONS

1. What is hydrogen peroxide?
2. How will you prepare this ear drop?
3. How will you calculate date of expiry?
4. What is the use of ear drop?
5. Is it a external preparation or internal preparation? Explain with reason.

EXERCISE NO. 4

Object : To prepare Boric acid ear drops

Requirements : Boric acid, denatured spirit, purified water, etc.

Theory : Boric acid is produced from native borax or from the other borates by reacting with hydrochloric acid or suphuric acid. It is colorless scales of a somewhat pearly luster or crystals but more commonly a white powder slightly unctous to the touch. It is odorless and stable in air and volatilizes with steam. It is slightly soluble in water hence in this preparation denatured spirit is used as it enhances the solubility of boric acid in purified water. It is a very weak germicide. It is a nonirritating hence suitable for application of delicate structures such as cornea of the eye.

Formula

Boric acid	0.2 g
Specially denatured spirit	3.0 ml
Purified water, sufficient to produce	15.0 ml

Procedure

1. Weighed quantity of boric acid is to be dissolved in denatured spirit.
2. Add purified water and make sufficient volume 15.0 ml.
3. If any impurities filter it and dispensed in a suitable container.

Category : Local anti-infective

Dose : 3 to 5 drops

Storage : Preserve in a well-closed container. The label on the container bears a warning that it is not for internal use. It should not be applied to extensive areas of broken skin.

Uses : Local anti-infective (germicide)

VIVA VOCE QUESTIONS

1. What is boric acid?
2. What is germicide explained with examples?
3. How ear drops are prepared?
4. Why it is suitable for eye?
5. Give storage and labeling requirement.

EXERCISE NO. 5

Object : To prepare Chloramphenicol ear drops

Requirements : Chloramphenicol, propylene glycol, etc.

Formula

Chloramphenicol	0.75 g
Propylene glycol, sufficient to produce	15.0 ml

Procedure

1. Dissolve Chloramphenicol in propylene glycol.
2. Shake and add remaining quantity of propylene glycol to produce sufficient volume of 15 ml.

Category : Antibiotic

Dose : 3 to 5 drops three to four times a day

Storage : Preserve in a well-closed container and protect from light.

Uses : Antibiotic, germicide, anti-microbial

VIVA VOCE QUESTIONS

1. What is antibiotic?
2. What is chloramphenicol?
3. What is the difference between broad spectrum and narrow spectrum antibiotics?
4. Give method of preparation of ear drop.
5. Give storage condition for ear drops.

EXERCISE NO. 6

Object : To prepare Framycetin ear drops

Requirements : Framycetin sulphate, water for injection, sterilized glycerin, etc.

Formula

Framycetin sulphate	0.3 g
Water for injection	0.6 ml
Sterilized glycerin, sufficient to produce	15.0 ml

Procedure

1. Dissolve framycetin sulphate in glycerin with water for injection.
2. Add sterilized glycerin to produce 15 ml.

Category : Antibiotic **Dose :** 3 to 5 drops

Storage : Preserve in a well-closed container. **Uses :** Germicide

VIVA VOCE QUESTIONS

1. What is framycetin?
2. How will you prepare ear drops?
3. Give pharmaceutical application of framycetin.

EXERCISE NO. 7

Object : To prepare Ichthammol ear drops

Requirements : Ichthammol, glycerin, etc.

Theory : Ichthammol consists of the ammonium salts of the sulphuric acids of an oily substance prepared from a bituminous schist or shale, or from other sources, together with ammonium sulphate and water. It is black viscid liquid, which has characteristic strong odor It is miscible in water, alcohol, glycerin and with fixed oils. It is used as antiparasitic and irritant,

Formula

Ichthammol	1.5 g
Glycerin, sufficient to produce	15.0 ml

Procedure : Same ad Exercise No. 6

Category : Pharmaceutical aid **Dose :** 3 to 5 drops

Storage : Preserve in a well-closed container. **Uses :** Antiparasitic

VIVA VOCE QUESTIONS

1. What is Ichthammol?
2. What is the use of glycerin in ear drops?
3. What is antiparasitic?

EXERCISE NO. 8

Object : To prepare Phenol ear drops

Requirements : Phenol glycerin, glycerin, etc.

Formula

Phenol glycerin	6.0 ml
Glycerin, sufficient to produce	15.0 ml

Procedure

1. Mix phenol glycerin with glycerin and shake it.
2. Add glycerin sufficient to produce 15 ml.

Category : Disinfectant

Dose : 3 to 5 drops

Storage : Preserve in well-closed container, Caution : These ear drops must not be diluted with water.

Uses : Disinfectant

VIVA VOCE QUESTIONS

1. What is phenol glycerin?
2. What is the use of phenol glycerin?
3. Why glycerin is used in this preparation?
4. Define disinfectant.
5. What is the difference between disinfectant and antiseptic?

EXERCISE NO. 9

Object : To prepare Phenol and Amethocaine ear drops

Requirements : Phenol glycerin, amethocaine hydrochloride, glycerin, etc.

Formula

Phenol glycerin	3.0 ml
Amethocaine hydrochloride,	0.15 g
Glycerin, sufficient to produce	15.0 ml

Procedure

1. Dissolve amethocaine hydrochloride in small proportion of glycerin.
2. Mix phenol glycerin.
3. Add sufficient glycerin to produce 15 ml.

Category : Local anaesthetic and Disinfectant

Dose : 3 to 5 drops

Storage : Preserve in well-closed container, Caution : These ear drops must not be diluted with water.

Uses : Local anaesthetic and Disinfectant

VIVA VOCE QUESTIONS

1. What is amethocaine?
2. How ear drops are prepared?
3. Give pharmaceutical application of ear drops.
4. Why this ear drop is not diluted?

EXERCISE NO. 10

Object : To prepare Hydrocortisone and Neomycin ear drops

Requirements : Neomycin sulphate, Hydrocortisone acetate, purified water, etc.

Formula

Neomycin sulphate	225 mg
Hydrocortisone acetate	75.0 mg
Purified water, sufficient to produce	15.0 ml

Procedure

1. Dissolve neomycin sulphate and hydrocortisone acetate with purified water.
2. Add sufficient quantity of water to produce 15 ml.

Category : Antibiotic

Dose : 3 to 5 drops

Storage : Preserve in well-closed container.

Uses : Antibiotic, anti-bacterial

VIVA VOCE QUESTIONS

1. What is antibiotic?
2. Give pharmaceutical application of ear drops

EXERCISE NO. 11

Object : To prepare Glyceric magnesium sulphate ear drops

Requirements : Magnesium sulphate, purified water, glycerin, etc.

Formula

Magnesium sulphate	1.5 g
Purified water	6.0 ml
Glycerin, sufficient to produce	15.0 ml

Procedure

1. Dissolve magnesium sulphate in purified water.
2. Add glycerin sufficient to produce 15 ml.

Category : Soften ear wax

Dose : 3 to 4 drops

Storage : Preserve in a well-closed container.

Uses : It is used to soften ear wax.

VIVA VOCE QUESTIONS

1. What is the use of magnesium sulphate in ear drops?
2. How it softens ear wax?
3. What is the ear wax?
4. How to prepare ear drops?

EXERCISE NO. 12

Object : To prepare Salicylic acid ear drops

Requirements : Salicylic acid, glycerin, denatured spirit, etc.

Formula

Salicylic acid	0.12 g
Glycerin	8.0 ml
Specially denatured spirit, sufficient to produce	15.0 ml

Procedure

1. Weighed amount of salicylic acid is to be dissolved in denatured spirit.
2. Add glycerin with shaking
3. Add purified water to produce 15 ml.

Category : Analgesic

Dose : 3 to 5 drops

Storage : Preserve in a well-closed container.

Uses : Analgesic

VIVA VOCE QUESTIONS

1. What is salicylic acid?
2. How will you prepare salicylic acid ear drops?
3. Give storage condition of ear drops.
4. What is the use of glycerin in this preparation?

EXERCISE NO. 13

Object : To prepare Sulphacetamide ear drops

Requirements : Sulphacetamide sodium, sodium metabisulphite, phenyl mercuric nitrate or acetate, purified water, etc.

Formula

Sulphacetamide sodium	4.5 g
Sodium metabisulphite	0.1%
Phenyl mercuric nitrate or acetate,	0.002%
Purified water, sufficient to produce	15.0 ml

Procedure

1. Dissolve Sulphacetamide sodium in purified water.
2. Dissolve sodium metabisulphite and phenyl mercuric nitrate in small amount of purified water.
3. Mix these two solution and shake.
4. Add sufficient purified water to produce 15 ml

Category : Sulphonamide

Dose : 3 to 5 drops

Storage : Preserve in a well closed container. It should be used within 4 weeks after opening the containers.

Uses : Sulphonamide, antibacterial

VIVA VOCE QUESTIONS

1. What is sulphonamide?
2. What is the difference between antibiotic and sulphonamide?
3. Give method of preparation of ear drops.
4. What is the use of sodium metabisulphite in this preparation?

5

Elixirs

- Low alcohol elixir
- High alcohol elixir
- Simple elixir
- Benzaldehyde elixir
- Paediatric chloral elixir
- Ephedrine hydrochloride elixir
- Paediatric paracetamol elixir
- Piperazine citrate elixir
- Phenobarbitone elixir
- Terpin hydrate elixir
- Elixir of vitriol

Elixirs are clear, liquid, flavored hydroalcoholic preparation intended for oral use. They contain one or more medicaments, pleasantly flavored, usually attractive color containing high proportion of alcohol or sucrose or suitable polyhydric alcohols together with suitable additives (including antimicrobial agents). The alcoholic contents in elixir vary from 5% to 40%. In general they are more stable than mixture as sufficient alcohol is added to maintain the drug in solution but it is prepare by dissolving granules or powder in water just dispersing patient for use. Most of the elixirs become turbid when moderately diluted by aqueous fluids.

Elixirs are mainly of two types

1. Non-medicated elixir : They are used purely as diluting agents or solvents for drugs containing approximately 25 percent alcohol, e.g. simple elixir, Iso-alcoholic elixir or low alcohol elixir (containing 8-10% alcohol), High alcoholic elixir (containing 75-78% alcohol)
2. Medicated elixirs : Elixirs containing therapeutically active compounds are known as medicated elixirs. e.g. phenobarbital elixirUSP, dexamethasone elixir USP etc.

The elixir is used as a vehicle for other drug in many commercially available cough syrups viz. Dextromethorphan hydrobromide, codeine phosphate, chlorahydrate maleate, pyrilamine maleate, ammonium chloride, creosote chloroform, and a wide variety of other drugs with expectorant and anti-tussive properties.

Some proprietary elixirs (phenethicillin and phenoxymethyl penicillin) are available in the market in granule or powder form because active ingredients are unstable in solution. They are dispersed by adding measured amount of water or at a specified point in bottle and shake until solution is complete. The preparation is labeled as to be stored in cool place and use within one week.

FORMULATION

Generally elixirs contain following ingredients.

Vehicles : About 10% to 20% of alcohol is used for keeping oils, vegetable extracts, tannin etc. in solution form. Faint opalescence from flavoring agents containing essential oils and light precipitants from vegetable extracts are not considered acceptable in elixirs. Glycerol and propylene glycol are used as solvents.

Stabilizers : In neomycin elixir, citric acid is used to adjust pH 4.0 to 5.0 to minimize the darkening that occurs on storage. Disodium editate should be used to sequester heavy metals that catalyse decomposition of the antibiotics.

Colouring agents : Elixirs are attractive preparation containing following dyes.

Solution	Color	Elixir
Amaranth	Magnetic red	Paediatric paracetamol, Paediatric streptomycin
Compound tartrazine	Saffron	Ephedrine, Isoniazid, Neomycin, Phenobarbitone
Green S, tartrazine	Green	Piperazine citrate

Flavouring agents : Sweeting agents and fruit flavors are used in many medicinal preparations etc. Black current syrup in chloral elixir, Concentrated raspberry juice with inverted syrup in paracetamol elixir, Lemon spirit with invert syrup in ephedrine elixir, compound orange spirits with glycerol in phenobarbital elixir.

Preservatives : In elixir, fermentation & mold growth are inhibited when it contains more than 20% of alcohol, propylene glycol or glycerol. Syrup containing preparation is also inhibitory to many microorganisms because of high osmotic pressure. The common used preservatives are double strength chloroform, spirit, and benzoic acid and methyl ester of parahydroxy benzoic acid.

METHOD OF PREPARATION

Elixirs are prepared by dissolving the ingredients in the suitable solvent. Usually, the alcohol soluble substances are dissolved in alcohol and water soluble in water separately. The aqueous solution is then added to the alcoholic solution with constant stirring.

CONTAINER

It should be stored in narrow mouthed screw capped colorless plain container.

LABELLING

It should comply with the general requirements for labelling, if the elixir is supplied as granules or powder to be reconstituted immediately before issue, the label on the container should mention-

1. Granules or powder for elixir
2. The nature and quantity of liquid to be used in constituting elixir.
3. Name and proportion of medicaments in the constituted product prepared in accordance with the manufacture is recommendations.
4. Condition under which the preparation should be stored.
5. Expiry date.
6. The period which the constituted product may be expected to retain its potency when prepared in accordance with the manufactures.

EXERCISE NO. 1

Object : To prepare Low alcohol elixir

Requirements : Compound orange spirit, alcohol, glycerin, sucrose, purified water, etc.

Formula

Compound orange spirit	1.0 ml
Alcohol	10.0 ml
Glycerin	20.0 ml
Sucrose	32.0 ml
Purified water, sufficient to produce	100.0 ml

Procedure

1. Prepare compound orange spirit.
2. Mix the alcohol, glycerin and 50.0 ml of purified water.
3. Add measured amount of compound orange spirit.
4. Agitate thoroughly from time to time.
5. Allow standing for 24 hours.
6. Filter this solvent mixture through hard filter papers.
7. Dissolve the weighed amount of sucrose in the filtrate by agitation or percolation.
8. Add enough of the solvent mixture to make sufficient volume 100.0 ml.

Category : Pharmaceutical aid and flavoring agent

Storage : Preserve in a well-closed container, protect from light

Uses : Pharmaceutical aid and flavoring agent

VIVA VOCE QUESTIONS

1. What is low alcohol elixir?
2. Why high alcoholic elixirs are required?
3. How to prepare low alcoholic elixir?
4. How to prepare orange syrup?
5. Give pharmaceutical uses of elixir.

EXERCISE NO. 2

Object : To prepare High alcohol elixir

Requirements : Compound orange spirit, saccharin, glycerin, alcohol, etc.

Formula

Compound orange spirit	0.40 ml
Saccharin	0.30 g
Glycerin	20.0 ml
Alcohol, sufficient to produce	100.0 ml

Procedure

1. Dissolve the compound orange spirit and the saccharin in 70.0 ml of alcohol.
2. Add measured amount of glycerin and mix.
3. Add sufficient amount of alcohol to produce 100.0 ml and mix properly.
4. Filter the mixture and preserve in suitable container.

Category : Pharmaceutical aid and flavoring agent

Storage : Preserve in a well-closed container, protected from light.

Uses : Pharmaceutical aid and flavoring agent

VIVA VOCE QUESTIONS

1. What is high alcohol elixir?
2. How will you prepare low alcoholic elixir?
3. What is the use of glycerin?
4. Why saccharin is used in place of sugar?
5. Give pharmaceutical uses of elixir.

EXERCISE NO. 3

Object : To prepare Paediatric chloral hydrate elixir

Requirements : Chloral hydrate, purified water, black current syrup, syrup, etc.

Theory : Chloral hydrate, chemically is 2,2,2 trichloro ethane-1,1-diol[$CCl_3CH(OH)_2$]. It is a colorless, transparent or white crystalline compound. It has slightly acrid odor and has bitter taste. It is widely used as a hypnotic that is more effective for inducing sleep than prolonging it. Somnifacient, doses promptly produce drowsiness and sedation, followed by quiet sound sleep within an hour.

It is also used in combination with barbiturates in the first stage of labour. It is used in the treatment of eclampsia and tetanus. In postoperative care it has been a valuable adjunct to opiate and analgesics. It has little or no analgesic action and should not be used alone in the presence of pain. Oral administration of chloral hydrate is rapidly converts it into trichloroethanol which is largely responsible for its hypnotic action.

Formula

Chloral hydrate	4.0 g
Black current syrup	20.0 ml
Purified water	2.0 ml
Syrup, sufficient to produce	100.0 ml

Procedure : Same as Exercise No. 2

Category : Sedative **Weight per ml :** 1.320 to 1.330 g

Dose : For a child up to 1 year, 2.5 to 5.0 ml. When a dose is less or not a multiple of 5.0 ml is prescribed, the elixir may be diluted with syrup. The diluted syrup must be freshly prepared.

Storage : Preserve in a well-closed container.

Uses : Sedative

VIVA VOCE QUESTIONS

1. What is chloral hydrate?
2. What is the difference between sedative and hypnotic?
3. What is current syrup?
4. How will you prepare elixir?

EXERCISE NO. 4

Object : To prepare Ephedrine hydrochloride elixir

Requirements : Ephedrine hydrochloride, glycerin, compound tartrazine solution, alcohol, chloroform spirit, lemon spirit, invert syrup, purified water, syrup, etc.

Formula

Ephedrine hydrochloride	0.3 g
Glycerin	20.0 ml
Compound tartrazine solution	1.0 ml
Alcohol (90%)	10.0 ml
Chloroform spirit	4.0 ml
Lemon spirit	0.02 ml
Invert syrup	20.0 ml
Purified water	6.0 ml
Syrup, sufficient to produce	100.0 ml

Procedure

1. Dissolve ephedrine hydrochloride in purified water.
2. Add gradually all additives with mixing.
3. Add sufficient quantity of syrup to produce 100 ml

Category : Sympathomimetic **Weight per ml :** 1.230 to 1.240 g

Dose : For an adult, 5 to 10 ml. For a child of age up to one year 2.5 ml. One to five years 6.0 ml, six to twelve year 10.0 ml. When a dose that is less than or not a multiple of 5.0 ml is prescribed, the elixir may be diluted with syrup.

Storage : Preserve in a well-closed container, protected from light.

Uses : Expectorant , bronchodialtor

VIVA VOCE QUESTIONS

1. How to prepare ephedrine hydrochloride elixir?
2. What is the use of individual additives in the elixir?
3. What is the use of glycerin?
4. What is the use of chloroform water?
5. Give storage conditions of elixir.
6. What is expectorant?

EXERCISE NO. 5

Object : To prepare Simple elixir

Requirements : Orange tincture, syrup, chloroform water, etc.

Formula

Orange tincture	7.5 ml
Syrup	40.0 ml
Chloroform water, sufficient to produce	100.0 ml

Procedure

1. Mix orange tincture with the syrup.
2. Add sufficient chloroform water to produce 100.0 ml.
3. Add 5.0 g of purified talc and shake vigorously.
4. Stand for a few hours and shake occasionally.
5. Filter the elixir and preserve in a suitable container.

Category : Pharmaceutical aid

Dose : 4 to 8 ml

Storage : Preserve in a well-closed container.

Uses : Pharmaceutical aid

VIVA VOCE QUESTIONS

1. What is elixir?
2. What is the difference between elixir and syrup?
3. How will you prepare orange syrup?
4. Give storage condition of elixir.
5. Why chloroform water is used in place of purified water?

EXERCISE NO. 6

Object : To prepare Diphenhydramine elixir

Requirements : Diphenhydramine hydrochloride, orange oil, cinnamon oil, clove oil, coriander oil, anethole, amaranth solution, alcohol, syrup, purified water etc,

Formula

Diphenhydramine hydrochloride	0.250 g
Cinnamon oil	0.010 ml
Clove oil	0.008 ml
Orange oil	0.025 ml
Coriander oil	0.003 ml
Anethole	0.003 ml
Amaranth solution	0.160 ml
Alcohol	15.0 ml
Syrup	35.0 ml
Purified water, sufficient to produce	100.0 ml

Procedure

1. Dissolve diphenhydramine hydrochloride, orange oil, cinnamon oil, clove oil, coriander oil in alcohol.
2. Add anethole and amaranth solution and mix it.
3. Add syrup and shake.
4. Add sufficient purified water to produce 100.0 ml.
5. If necessary filter the elixir.

Category : Anti-histaminic **Dose :** 10 to 20 ml, upto four times a day.

Storage : Preserve in a well closed container, protect from light.

Uses : Anti-histaminic

VIVA VOCE QUESTIONS

1. What is diphenhydramine?
2. Why elixir is required?
3. Give storage condition of elixir?
4. Why three oil different are used ?
5. Define the term antihistamine.

EXERCISE NO. 7

Object : To prepare Paediatric paracetamol elixir

Requirements : Paracetamol, ethanol (96%), Propylene glycol, chloroform spirit, concentrated raspberry juice, Invert syrup, Amaranth solution, Glycerin, etc.

Formula

Paracetamol	2.4 g
Ethanol (96%)	10.0 ml
Propylene glycol	10.0 ml
Concentrated raspberry juice	2.5 ml
Chloroform spirit	2.0 ml
Invert syrup	27.5 ml
Amaranth solution	0.2 ml
Glycerin, sufficient to produce	100.0 ml

Procedure

1. Mix ethanol (96%), propylene glycol and chloroform spirit and make a mixture.
2. Dissolve paracetamol and shake it.
3. Dilute concentrated raspberry juice with invert syrup.
4. Add diluted solution of raspberry juice in paracetamol mixture.
5. Add amaranth solution and shake.
6. Add sufficient amount of glycerin to produce 100 ml.

Category : Analgesic and antipyretic **Weight per ml :** 1.210 to 1.230 g

Dose : 5 to 10 ml according to the age of patient. Paediatric paracetamol elixir should be diluted.

Storage : Preserve in a well-closed container and protect from light.

Uses : Analgesic and antipyretic

VIVA VOCE QUESTIONS

1. What is paracetamol?
2. How will you prepare paracetamol elixir?
3. Define analgesic and antipyretic activity.
4. What is the use of propylene glycol?
5. Give pharmaceutical application of elixir.

EXERCISE NO. 8

Object : To prepare Piperazine citrate elixir

Requirements : Piperazine citrate, chloroform spirit, glycerin, orange oil, syrup, purified water, etc

Formula

Piperazine citrate	18.0 g
Chloroform spirit	0.5 ml
Glycerin	10.0 ml
Orange oil	0.025 ml
Syrup	50.0 ml
Purified water, sufficient to produce	100.0 ml

Procedure

1. Dissolve piperazine citrate in required amount of the purified water.
2. Add gradually orange oil, glycerin, syrup and chloroform spirit and mix.
3. Add sufficient purified water to produce 100.0 ml.

Category : Expectorant

Dose : For threadworms – 4 to 5 ml daily in divided doses. For roundworms as a single dose up to 30 ml, according to the age of the patient.

Storage : Preserve in well-closed container, protected from light.

Uses : Expectorant

VIVA VOCE QUESTIONS

1. What is elixir?
2. What is threadworm and roundworm?
3. Why a single dose therapy is used for roundworm infection?
4. How elixirs are prepared?
5. What is the use of glycerin?
6. What is hygroscope? Explain with examples.

EXERCISE NO. 9

Object : To prepare Phenobarbitone elixir

Requirements : Phenobarbitone, alcohol, compound orange spirit, glycerol, compound tartrazine solution, purified water, etc.

Formula

Phenobarbitone	0.3 g
Alcohol (90%)	40.0 ml
Compound orange spirit	2.4 ml
Glycerol	40.0 ml
Compound tartrazine solution	1.0 ml
Purified water, sufficient to produce	100.0 ml

Procedure

1. Dissolve phenobarbital in the alcohol (90%).
2. Add compound orange spirit, glycerin and compound tartrazine solution.
3. Add sufficient water to produce 100 ml.
4. Add 2.5 g of purified talc, previously sterilized and shake for a few minutes.
5. Allow to stand for a few hours and shake occasionally.
6. Filter the elixir and dispensed in the suitable container.

Category : Anti-convulsant and hypnotic

Weight per ml : 1.045 to 1.055 g

Dose : 15 ml daily, increasing to 100 ml daily, in accordance with the needs and response of the patient. When a dose is less than or not a multiple of 5.0 ml is prescribed the elixir may be diluted with syrup. The diluted elixir must be freshly prepared.

Storage : Preserve in a well-closed container, protected from light.

Uses : Anti-convulsant and hypnotic

VIVA VOCE QUESTIONS

1. What is elixir?
2. What is the difference between elixir and syrup?
3. What is the use of phenobarbital?
4. What is the difference between hypnotic and sedative?

EXERCISE NO. 10

Object : To prepare Terpin hydrate elixir

Requirements : Terpin hydrate, orange oil, glycerin, alcohol, syrup, purified water, etc.

Formula

Terpin hydrate	5.0 g
Orange oil	0.02 ml
Glycerin	40.0 ml
Alcohol	42.5 ml
Syrup	10.0 ml
Purified water sufficient to produce	100.0 ml

Procedure

1. Dissolve terpin hydrate in alcohol.
2. Add successively the orange oil, glycerin, syrup, and mix.
3. Add sufficient purified water to produce 100 ml and mix.
4. If necessary, filter the elixir and preserve in a suitable container.

Category : Expectorant

Dose : 4.0 ml

Storage : Preserve in a well-closed container which prevents access of moisture.

Uses : Expectorant

VIVA VOCE QUESTIONS

1. What is terpin hydrate?
2. What is the use of terpin hydrate?
3. How to prepare elixir?
4. Give pharmaceutical application of elixir.

EXERCISE NO. 11

Object : To prepare Elixir of vitriol

Requirements : Strong ginger tincture, cinnamon oil, sulphuric acid, alcohol (90%), etc.

Formula

Strong ginger tincture	5.0 ml
Cinnamon oil	0.15 ml
Sulphuric acid	7.0 ml
Alcohol (90%), sufficient to produce	100.0 ml

Procedure

1. Mix the sulphuric acid gradually with 60 ml of alcohol(90%) and cool.
2. Dissolve the cinnamon oil in small proportion of alcohol.
3. Add the measured volume of strong ginger tincture.
4. Add sufficient alcohol to produce 100 ml.

Category : Pharmaceutical aid

Dose : 0.3 to 1.2 ml

Storage : Preserve in a well-closed container.

Uses : Pharmaceutical aid

VIVA VOCE QUESTIONS

1. Why elixirs are prepared?
2. What is the use of sulphuric acid in this preparation?
3. How will you preserve elixir?
4. How elixirs differ from syrup?

6

Enemas

- Chloral hydrate enema
- Magnisium sulphate enema
- Paraldehyde enema
- Soap enema
- Opium enema
- Phosphate enema

Enemas are aqueous or oily solutions or suspension that are employed to evacuate the bowel, to influence the general system by absorption or to affect locally the seat of disease of introduced into the rectum for cleansing, therapeutic or diagnostic purpose. Cleansing preparation are used to evacuate feces in constipation before an operation. Enemas are usually given at body temperature in quantities of 1 to 2 pt injected slowly with a syringe. If they have to be retained in the intestine they should not be used in larger quantities than 6 fl. ounces for an adult.

Sodium chloride, sodium bicarbonate, sodium monohydrogen phosphate and sodium dihydrogen phosphate are used in enemas. Following substances may be used alone or in combination with each other or in combination with irritants such as soap.

The action of enemas are classified as

1. By stimulation peristalsis because of their volume is large about 0.5 to 1.0 liter e.g. plain water, soap and turpentine enemas or other type of enemas causes osmotic retention of water in the bowel.
2. By lubricating impacted feces e.g. This type of enema is a smaller volume like olive and arachis oil enemas having volume about 100 to 500 ml.

Therapeutic enemas may be used as :

1. Anthelmintic : Quassia enema for thread worms
2. Sedatives : Chloralhydrate, paraldehyde enemas.
3. Anti-inflammatory agent : Corticosteroids for ulcerative colitis
4. Stimulating enemas
5. Nutritive : when absorption by mouth is impaired, enemas are used, e.g. barium sulphate suspension is used for X-ray examination of the lower bowel. They may contain of radio opaque substances for roentgenographic examination of the lower bowel.

CONTAINERS

Large volume enemas are administered from a douche can and should be warmed to body temperature before use. Some commercially available small volume enemas are in disposable polyethylene or polyvinyl chloride bags sealed with a rectal nozzle.

LABELLING

1. To be warmed to body temperature before use
2. 'Not to be taken orally'
3. 'For rectal use only'

STORAGE

The enema should be stored in a cool place in complete darkness and an expiry date of only two days after issue is advisable.

EXERCISE NO. 1

Object : To prepare Chloral hydrate enema

Requirements : Chloral hydrate, mucilage of starch, etc.

Formula

Chloral hydrate	2.0 g
Mucilage of starch, sufficient to produce	100.0 ml

Procedure

1. Dissolve weighed quantity of chloral hydrate in mucilage of starch.
2. Add sufficient to produce 100 ml.

Category : Sedatives **Dose :** 120 ml. by rectal injection.

Storage : Preserve in a well-closed container.

Uses : Sedatives

VIVA VOCE QUESTIONS

1. What is enema?
2. Why enema is required?
3. How to use enema?
4. What is sedative? Explain with example.

EXERCISE NO. 2

Object : To prepare Magnesium sulphate enema

Requirements : Magnesium sulphate, purified water, etc.

Formula

Magnesium sulphate	50.0 g
Purified water, sufficient to produce	100.0 ml

Procedure

1. Dissolve magnesium sulphate in purified water and shake.
2. Add sufficient purified water to produce 100 ml.
3. If, necessary, filter the enema.

Category : Saline laxative

Dose : 120 ml, by rectal injection.

Storage : Preserve in a well-closed container.

Uses : Saline laxative

VIVA VOCE QUESTIONS

1. What is the difference between enema and douche?
2. What is the use of magnesium sulphate enema?
3. How to use enema?
4. What is the utility of enema?

EXERCISE NO. 3

Object : To prepare Paraldehyde enema

Requirements : Paraldehyde, sodium chloride, purified water, etc.

Formula

Paraldehyde	8.0 ml
Sodium chloride	0.9 g
Purified water, sufficient to produce	100.0 ml

Procedure

1. Dissolve sodium chloride in purified water and shake it and filter.
2. Add paraldehyde and dissolve.
3. Add sufficient purified water to produce 100 ml.

Category : Sedatives

Dose : 40 to 50 ml. equivalent to 4 to 5 ml. of paraldehyde per kg. of body weight.

Storage : Preserve in a well-closed container.

Uses : Sedatives

VIVA VOCE QUESTIONS

1. What is the use of paraldehyde enema?
2. How to prepare enema?

EXERCISE NO. 4

Object : To prepare Phosphate enema

Requirements : Sodium acid phosphate, sodium phosphate, purified water, etc.

Formula

Sodium acid phosphate	160 g
Sodium phosphate	60.0 g
Purified water, sufficient to produce	100.0 ml

Procedure

1. Mix sodium acid phosphate and sodium phosphate and powder in pestle mortar.
2. Dissolve this mixture in purified water.
3. Add sufficient purified water to produce 100 ml.

Category : Saline laxative.

Dose : 100 ml

Storage : Preserve in a well-closed container.

Uses : Saline laxative.

VIVA VOCE QUESTIONS

1. What is the use of sodium acid phosphate?
2. How enemas are prepared?
3. What is laxative?

EXERCISE NO. 5

Object : To prepare Soap enema

Requirements : Soft soap, purified water, etc

Formula

Soft soap	5.0 g
Purified water(lukewarm), sufficient to produce	100.0 ml

Procedure

1. Dissolve weighed amount of soft soap in warm purified water.
2. Add sufficient warm purified water to produce 100.0 ml.

Dose : 500 ml by rectal injection.

Storage : Preserve in a well-closed container.

Uses : To be warmed to body temperature before use.

VIVA VOCE QUESTIONS

1. What is the use of soft soap?
2. Give storage condition of enema.
3. How to use enema?

EXERCISE NO. 6

Object : To prepare Opium enema

Requirements : Opium tincture, mucilage of starch, etc.

Formula

Opium tincture	3.3 ml
Mucilage of starch, sufficient to produce	100.0 ml

Procedure

1. Prepare mucilage of starch and add opium tincture.
2. Add sufficient starch mucilage to produce 100 ml.

Category : Hypnotic and sedative

Storage : Preserve in a well-closed container.

Uses : Hypnotic and sedative

VIVA VOCE QUESTIONS

1. What is the use of opium enema?
2. What is tincture?
3. What is mucilage?
4. What is difference between hypnotic and sedative agent?

7

Emulsions

- Liquid paraffin emulsion
- Castor oil emulsion
- Liquid paraffin and Magnesium hydroxide emulsion
- Concentrated peppermint emulsion
- Oral sulfadiazine emulsion
- Fat emulsion for parenteral nutrition

Emulsion is a heterogeneous system consisting of at least one immiscible liquid dispersed in another in the form of droplets whose diameter in general exceed 0.1μm. Such system posses a minimal stability because the droplets quickly coalesces and the two liquids gets separated. The stability of the emulsion is increased by adding another substance known as emulsifying agent or emulsifiers. The liquid droplets, generally known as the emulsion phase or emulsion globules or dispersed phase or internal phase or the interrupted phase while the liquid in which they are dispersed is known as the continuous phase or the dispersion medium or the external phase or the continuous phase.

In other words an emulsion may, therefore be defined as a disperse system consisting of two immiscible liquids one of which is distributed in small globules throughout the other, the system being stabilized by the presence of a third substance called the emulsifying agent. e.g. milk is the ideal and natural emulsion.

PURPOSE OF EMULSION

1. It increases stability of many drugs, which are unstable in aqueous solution.
2. It improves the taste of objectionable medicinal agents and makes them more acceptable form.
3. It improves appearance of those material which are generally intended for topical application.

4. It improves penetration and spreading ability.
5. It enhances the rate and extent of absorption through alimentary canal.
6. It can be used as diagnostic agent. e.g. radiopaque emulsion in X-ray examination.
7. It improves solubility of many drugs which have limited solubility in aqueous phase.
8. It can prolong drug action.

EMULSION TYPES

1. Water-in-oil (w/o) emulsion
   Dispersed phase – water
   Continuous phase – oil
2. Oil-in water (o/w) emulsion
   Dispersed phase – oil
   Continuous phase – water
3. Multiple emulsion
   (a) Water-in-oil-in-water (w/o/w)
   (b) Oil-in water-in oil (o/w/o)
4. Miscellaneous
   Globule size or the disperse phase size of emulsion can vary enormously but are commonly in the range of 0.25-25μm diameter. On the basis of size emulsions are mainly of three types
   a. Coarse emulsion : globule size more than 25.0 μm.
   b. Fine emulsion : globule size less than 5.0 μm.
   c. Microemulsion of miceller emulsion : globule size range of 10-75 nm such emulsions appears transparent to the human eye in daylight.

EMULSIFYING AGENTS

An emulsifying agent must be employed in the preparation of stable emulsion. Its purpose is to prevent coalescence of droplets and maintain the stability of the individual droplets of the continuous phase. Any compounds that lower interfacial tension and forms a film at an interface can act as an emulsifying agent. The effectiveness of an emulsifying agent depends on the following factors viz.

1. Chemical structure
2. Concentration
3. Solubility
4. pH
5. Physical properties
6. Electrostatic effect

Emulsifying agents are classified as :

I. HYDROCOLLOIDS

A. ***Natural products***

a. Tree exudate : gum arabic (acacia), gum ghatti, karaya gum, tragacanth
b. Sea weeds : agar, carrageenan, alginate etc.
c. Seed extracts : locust bean, guar gum, quince seed
d. Steroid containing substance : beeswax, woolfat, wool alcohol
e. Animal substances : gelatin, casein etc.
f. Mineral substances : bentonite, veegum

B. ***Inorganic products*** **:** Colloidal alumia, milk of magnesia, mag. oxide, magnesium trisilicate etc.

C. ***Semisynthetic products*** **:** methyl cellulose, carboxy methyl cellulose, hydroxy methyl cellulose, microcystalline cellulose.

D. ***Synthetic products*** **:** carbopol, colloidal silicone dioxide.

II. SURFACTANTS

A. ***Anionic group***

a. Carboxylic acid : soap, lactylates, polypeptide condensates
b. Sulfuric acid esters : sulfated monoglycerides, alkyl sulfates
c. Alkyl and alkyl-aryl sulfonates : dodecylbenzene sulfonates
d. Phosphoric acid esters : trioleyl phosphate
e. Substituted alkyl amides : sarcosinates, taurates
f. Hemiesters : sulfosuccinates

B. ***Cationic group***

a. Amines : alkoxyalkylamines,
b. Quaternaries : benzylkonium chloride

C. ***Nonionic group***

a. Polyalkoxyethers : polyoxyethylene alkyl/aryl ethers, polyoxyethylene polyoxypropylene, block polymers.Polyalkoxyesters : polyoxyethylene fatty acid esters, polyoxyethylene sorbiton acid esters
b. Polyalkoxyamides fatty acid esters : sorbitan esters, glyceryl esters, sucrose esters
c. Fatty alcohols : lauryl alcohol.

METHOD OF PREPARATION

A. ***Wet gum method*** **:** A primary emulsion can be prepared by various methods. However it depends on type of ingredients involved in mainly stable preparation of fixed oil, water and acacia (ratio in parts 4:2:1) is prepared by the following methods.

(a) Two parts of water and one parts of acacia mixture is to be triturated with a mortar and pestle until a smooth mucilage is formed.

(b) Oil is to be added slowly with continuous triturating until a smooth cream of primary emulsion is formed.

(c) The mixture should be again triturated for another 5 minutes and then add water to make up the volume with continuous triturating.

B. ***Dry gum method*** **:** A primary emulsions of fixed oil, water and acacia (ratio in parts 4:2:1) is prepared by the following methods.

(a) The oil is mixed in acacia with mortar and pestle until acacia powder is distributed uniformly.

(b) Purified water is to be added and rapidly trituration until to form the primary emulsion.

(c) Add other additives and remaining quantity of water is added with continuous trituration to finish the product.

C. ***Bottle method*** **:** It is a modified method of dry gum method. The ratio of oil, water and acacia should be 3:2:1 or 2:1:1 is used for the preparation of emulsion. As the low viscosity of the volatile oil requires a higher proportion of acid.

(a) The oil is mixed in acacia by shaking the bottle uniformly

(b) Add measured quantity of water and shake until uniform emulsion is formed.

D. ***Soap method*** **:** A soap is formed by mixing relatively equal volumes of an oil and an aqueous solution containing a sufficient amount of alkali. This method is suitable for the formation of an o/w or w/o emulsion. It is dependent on the formation of soap. e.g. olive oil and lime water are mixed during the preparation of calamine lotion to form calcium oleate, an emulsifying agents.

ADDITIVES

1. Thickening agents : e.g. emulsifying agent
2. Antimicrobial agents

(a) Acids and acid derivatives : benzoic acid, sorbic acid, propionic acid, dehydro-acetic acid

(b) Alcohols : chlorobutanol, phenoxy-2-ethanol

(c) Aldehydes : Formaldehyde, glutareldehyde
(d) Phenolics : phenol, cresol, chlorothymol, o-phenylphenol, p-chlorometaxylenol, methyl p-hydroxybenzoate, propyl p- hydroxybenzoate, butyl p- hydroxybenzoate
(e) Quaternaries : benzethonium chloride, benzylkonium chloride, cetylpridinium chloride, cetyltrimethyl ammonium bromide, chlorhexidine and salts
(f) Mercurials : phenylmercuric acetate, sodium ethylmercurithiosalicylate
(g) Miscellaneous : 6-acetoxy-2,4-dimethyl-m-dioxane, Imidizolidinyl urea compound

3. Antioxidant : ascorbic acid, ascorbyl palmitate, butylated hydroxyanisol, butylated hydroxytoluene, gallic acid, 4-hydroxymethyl-2,6-di-tert-butylphenol, L-tocopherol, sulfites.
4. Organoleptic agent : e.g. coloring agent, flavoring agent, and sweetening agent

EVALUATION OF EMULSIONS

Evaluation should be evaluated for following parametrs.

1. Types of emulsion (o/w or w/o) : (a) Dilution test (b) Dye test (c) Fluorescence test
2. conductivity test (e) $Cocl_2$/filter paper test (f) Creaming test
3. Particle size or globule size analysis by microscopic method
4. Rate of phase separation by centrifugation
5. Viscosity by viscometer
6. Electrophoretic properties
7. Measurement of dielectric constant
8. Measurement of conductivity
9. Microwaves irradiation
10. Organeleptic prpperties

INSTABILITY OF EMULSION

A stable emulsion is one in which the globules retain their initial character like mean size and size distribution and remain uniformly distributed throughout the continuous phase. During storage of emulsions instability is endorsed by creaming, cracking, reversible aggregation and or irreversible aggregation etc.

Cracking or coalescence : Coalescence is a growth process during which the emulsified particles join to form large particles. When small droplets are merged and large droplets are formed that suggests, the emulsion will separate completely and can not be re-dispersed by shaking. This process is also known as cracking. Cracking may be caused by any chemical, physical and biological effects that changes the nature of the interfacial film of an emulsifying agent. Creaming can be reduced by the following methods.

1. Addition of an emulsifying agent of opposite type
2. Decomposition for precipitation of emulsifying agent
3. Addition of the common solvent
4. Addition of proper preservatives
5. Incorporation of excess of disperse phase

Creaming : Under the influence of gravity suspended particles or globules tend to upward movement is known as creaming while downward movement of particles or droplets as sedimentation. Creaming or sedimentation depend on the differences in specific gravity between the phases. If creaming take place without any aggregation, the emulsion can be reconstituted by shaking. Process of creaming is explained by Stoke's law.

$$v = \frac{d^2 (\rho_1 - \rho_2) g}{18 \eta}$$

where,
v = terminal or settling velocity
d = diameter of globules
ρ_1 = density of disperse phase
ρ_2 = density of continuous phase
η = viscosity of dispersion medium
g = acceleration due to gravity

Factors that influence the rate of creaming and sedimentation.

1. Globules size : Globules size are of small size tendency to form creaming is less.
2. Viscosity : Higher the viscosity of continuous phase, less creaming
3. Density : Difference in density of two phases if less, it states more stability of emulsion.
4. Temperature : Lower temperature is more suitable for the better stability of emulsion

Phase inversion : It was observed that w/o emulsion of benzene in water that were stabilized with sodium stearate invert to o/w emulsions upon heating and reform w/o emulsions upon cooling. The temperature at which the inversion occur depends on the emulsifier concentration and is called phase inversion temperature (PIT). An o/w emulsion stabilized by a non-ionic polyoxyethylene-derived surfactant contains oil-swollen misceller of the surfactant as well as emulsifying oil.

EQUIPMENTS

1. Kenwood Mixer
2. Silversion Mixer
3. The Q.P. emulsifiers
4. Colloid Mills
5. Standard slurry-type dispersed mixer with vaned-rotor
6. Standard paste-type dispersed mixer with clipped rotor
7. Two stage homogenizer

CONTAINERS

Emulsion should be supplied in wide mouthed bottles

STORAGE

Emulsion should be kept at temperature not exceeding 20°. They should not be allowed to freeze.

LABELLING

Comply with general requirements for labelling, the addition, the label indicated following information.

1. Quantity of emulsifying agent
2. Preservative
3. "Shake well before use"
4. Keep in cool place.

EXERCISE NO. 1

Object : To prepare Liquid Paraffin emulsion

Requirements : Liquid paraffin, indian gum in powder, tragacanth in powder,Sodium benzoate, vanillin, glycerin, chloroform, purified water, etc

Formula

Liquid paraffin	50.0 ml
Indian gum in powder	12.5 g
Tragacanth in powder	0.5 g
Sodium benzoate	0.5 g
Vanillin	0.05 g
Glycerin	12,5 ml
Chloroform	0.25 ml
Purified water, sufficient to produce	100.0 ml

Procedure

1. Triturate the weighed quantity of liquid paraffin and chloroform with the Indian gums, tragacanth the vanillin in mortar and pestle.
2. Add 25.0 ml of purified water and triturate until a creamy emulsion is formed.
3. Dissolve sodium benzoate in small quantity of purified water.
4. Add glycerin and solution of sodium benzoate with continuous triturate.
5. Add sufficient purified water to produce 100 ml of emulsion and mix properly.

Category : laxative **Dose :** 8 to 30 ml

Storage : Emulsion should be kept at temperature not exceeding 20°C. It should not be allowed to freeze.

Uses : laxative

VIVA VOCE QUESTIONS

1. What is emulsion?
2. What is the difference between emulsion and suspension?
3. How to prepare emulsion? Define primary emulsion.
4. Define laxative, purgative and cathartic effect.
5. Give application of liquid paraffin.
6. What is difference between o/w and w/o emulsion?
7. How to identify type of emulsion?

EXERCISE NO. 2

Object : To prepare Castor oil emulsion

Requirements : Castor oil, acacia in powder, cinnamon water, etc.

Formula

Castor oil	37.5 ml
Acacia in powder	10.0 g
Cinnamon water, sufficient to produce	100.0 ml

Procedure

1. Triturate the weighed quantity of castor oil and the cinnamon water with acacia in mortar and pestle.
2. Add remaining quantity of cinnamon water to produce 100 ml.

Category : Laxative

Dose : 30 to 60 ml

Storage : Preserve in a well-closed container.

Uses : Laxative

VIVA VOCE QUESTIONS

1. What is primary emulsion?
2. What is secondary emulsion?
3. How to prepare castor oil emulsion?
4. Define laxative with examples.
5. What is phase inversion?
6. What is the difference in creaming and cracking?

EXERCISE NO. 3

Object : To prepare Liquid paraffin and Magnesium hydroxide emulsion

Requirements : Liquid paraffin, chloroform spirit, magnesium sulphate, sodium hydroxide, light magnesium oxide, chloroform, purified water, etc.

Formula

Liquid paraffin	25.0 ml
Chloroform spirit	1.5 ml
Magnesium hydroxide mixture	100.0 ml

(A) Magnesium hydroxide mixture

Magnesium sulphate	4.75 g
Sodium hydroxide	1.5 g
Light magnesium oxide	5.25 g
Chloroform	0.25 ml
Purified water, sufficient to produce	100.0 ml

Procedure

1. Mix the chloroform spirit with 65 ml of magnesium hydroxide mixture.
2. Add liquid paraffin and mix.
3. Add sufficient magnesium hydroxide to produce 100 ml.
4. Pass through the homogeniser.

Category : Laxative and antacid

Dose : 5 to 20 ml.

Storage : Preserve in a well-closed container

Uses : Laxative and antacid

VIVA VOCE QUESTIONS

1. What is the use of liquid paraffin?
2. Why is the globule size of emulsion is an important factor in emulsion?
3. What is surfactant?
4. What is the use of emulsion?
5. What is antacid?

EXERCISE NO. 4

Object : To prepare Concentrated peppermint emulsion

Requirements : Peppermint oil, polysorbate-20, double strength chloroform water, purified water, etc.

Formula

Peppermint oil	2.0 ml
Polysorbate-20	0.10 ml
Double strength chloroform water,	50.0 ml
Purified water, sufficient to produce	100.0 ml

Procedure

1. Mix peppermint oil with the polysorbate-20 by shaking.
2. Polysorbate-20 add gradually with double strength chloroform water.
3. Add sufficient purified water to produce 100 ml with shaking.

Category : Carminative

Dose : 0.25 to 1.0 ml.

Storage : Preserve in a well closed container

Uses : Carminative

VIVA VOCE QUESTIONS

1. What is the use of peppermint emulsion?
2. What is the source of peppermint?
3. How to prepare peppermint emulsion?
4. What is the difference between o/w and w/o emulsion.
5. How to identify type of emulsion?
6. How to select emulsifying agent?

EXERCISE NO. 5

Object : To prepare Oral sulfadiazine emulsion (o/w)

Requirements : Cottonseed oil, sulfadiazine, sorbitan monostearate, polyoxyethylene sorbitan monostearate, sodium benzoate, sweetner, flavor oil, purified water, etc.

Theory : Sulfadiazine is essentially water insoluble. Hence its suspension or emulsion is required. For the preparation of stable sulphadiazine suspension viscosity of the final product must be reasonably high. This could be achieved by the use of gums or by developing an emulsion of high internal phase. The choice of a cottonseed oil emulsion is probably arbitrary except for the fact that o/w preparations are quite palatable.

The HLB value of cottonseed oil should be adjusted to get stable emulsion of 10.0. One emulsifier polyoxyethylene (4) sorbiatan monostearate is satisfactory because HLB is 9.6 but mixed emulsifiers are generally preferred because they are safe and economic. Sulfadiazine emulsion is prepared using two emulsifier polyoxyethylene(20) sorbiatan monostearate (HLB-14.9) and sorbiatan monostearate (HLB-4.7).

$$\text{HLB value} = a \times 4.7 + b \times 14.9 = \text{about to HLB value } 10.0$$

Where,
a - ratio of sorbiatan monostearate
b - ratio of polyoxyethylene (20) sorbiatan monostearate

For smooth product, it is necessary to reduce the particle size of the sulfadiazine. For this purpose, blend of drug, oil, and oleophilic emulsifier is warmed slightly and triturated by passed through a colloidal mill or by pestle mortar. The emulsion itself is formed by adding the drug suspension to the aqueous phase, but in contrast to usual emulsion technique, the two phases are blended at different temperatures. Heating of drug suspension to 65°C would materially lower its viscosity and cause excessive settling of the drug particles unless specialized stirring equipment were employed.

The addition of the flowering oils at lower temperature prevents loss due to volatility. The emulsion should be protected against molds and fungi by use of sodium benzoate for this purpose is of doubtful merit while sulfadiazine has a broad anti-microbial spectrum. The emulsion having no antioxidant, cottonseed oil as an additive partially act as an antioxidant.

Formula

Cottonseed oil	46.0 g
Sulfadiazine	20.0 g
Sorbitan monostearate	8.4 g
Polyoxyethylene sorbitan monostearate	3.60 g
Sodium benzoate	0.20 g
Sweeting agent	qs
Flavor oil	qs
Purified water, sufficient to produce	100.0 g

Procedure

1. Mix Cottonseed oil, sulfadiazine, and sorbitan monostearate and heat at 50°C, pass through colloidal mill.(A)
2. Mix , polyoxyethylene sorbitan monostearate, sodium benzoate, sweetner, and purified water and heat at 65°C. (B)
3. Add (A) to (B) and stir while cooling to 45°C.
4. Add flavoring oil in this emulsion and continue stirring until cooled at room temperature.

Category : Broad spectrum anti-microbial

Dose : Initial dose 3 g, subsequent doses, upto 4 g daily, in divided doses.

Storage : Preserve in a well-closed container, protect from light.

Uses : Broad spectrum anti-microbial

VIVA VOCE QUESTIONS

1. What is emulsion?
2. What is true emulsion, micro-emulsion and coarse emulsion?
3. Why cottonseed oil is used in this preparation?
4. What is antioxidant?
5. What is preservative and explain with examples?
6. What is broad spectrum anti-bacterial/
7. Give pharmaceutical application of emulsion.
8. How to determine the type of emulsion?

EXERCISE NO. 6

Object : To prepare Fat emulsion for Parenteral nutrition o/w

Requirements : Cottonseed oil, polythylene glycol 200 monopalmitate, tartaric acid ester of cotton seed fatty acid mono glyceride, Polyoxyethylene-polyoxypropylene block polymer, and isotonic glucose solution, etc.

Theory : This emulsion is unusual since it could be rarely used as pH of the emulsion is changed during storage by hydrolysis of lipid. For better stability of the emulsion, use of a system of Pluronic F 68 and L-α-phosphatidyl choline.

Formula

Cottonseed oil	15.0 g
Polythylene glycol 200 monopalmitate	1.2 g
Tartaric acid ester of cotton seed fatty acid mono glyceride	0.30 g
Polyoxyethylene-polyoxypropylene block polymer	0.3 g
Isotonic glucose solution	83.20 g

Procedure

1. Blend cottonseed oil, polythylene glycol 200 monopalmitate and tartaric acid ester of cotton seed fatty acid mono glyceride and heat at 45°C. (A)
2. Blend polyoxyethylene-polyoxypropylene block polymer, and isotonic glucose solution, heat at 45°C. (B)
3. Add (A) to (B) at 40°C to 45°C and raise pressure to about 4000 psi.
4. Continue homogenizing until emulsion has been cycled five times.
5. Pack and autoclave at 121°C for 17 min.

Category : Fat nutrient

Storage : Preserve in a well-closed container, protect from light.

Uses : Fat nutrient

VIVA VOCE QUESTIONS

1. What is fat emulsion?
2. How to sterilize the emulsion?
3. What is the use of cotton seed oil?
4. What is the emulsifier?

8

Linctuses

- Pholcodine linctus
- Codeine linctus
- Simple linctus
- Paediatric simple linctus
- Paediatric opiate squill linctus
- Paediatric compound tolu linctus
- Methadone linctus
- Noscapine linctus

Linctuses are solutions of one or more medicaments, usually containing large amounts of sucrose, and sometimes glycerin, as a sweeting agent and also has a demulcent effect on the mucous membrane of the throat. The vehicle syrup is soothing the sore mucous membrane of the throat, has a beneficial effect. Linctuses are usually intended for use in the treatment of cough, being sipped and swallowed slowly without the addition of water. For local and prolong action, they should be administered undilluted and sipped and swallowed slowly. The usual dose of linctus is 5.0 ml.

Dilution of Linctuses : If a dose is prescribed which is less than or not a multiple of 5 ml, the linctus should be suitably diluted to achieve a dose-volume that is a multiple of 5.0 ml, using the specified diluent. They are diluted by syrup except codeine linctus, diabetic. It is diluted by chloroform water. For this purpose chloroform water must be freshly prepared. Diluted Linctuses should not be used within two weeks of preparation, or any other period as may be specified as may be authorized, and the date after which they should not be used is stated on the label. They contain following substances.

1. Vehicles : Syrup, tolu syrup, Invert syrup, glycerin, chloroform water, sorbitol etc.
2. Stabilizers : Syrups
3. Coloring agents : Compound tartrazine solution, coal tar dyes, etc.
4. Flavoring agents : Tolu syrup, fruit flavored syrup, lemon syrup, blackcurrent syrup.
5. Preservatives : Benzoic acid, chloroform spirits, cinnamic acid, tolu syrups etc.

e.g. codeine linctus, methadone linctus, padiatric opiate squill linctus, pholcodine linctus simple linctus, paediatric linctus, opiate squill linctus, noscapine linctus, etc.

LABELLING

50 ml. to be dispensed unless otherwise directed. They should be sipped and swallowed slowly without addition of water.

EXERCISE NO. 1

Object : To prepare Codeine linctus

Requirements : Codeine phosphate, benzoic acid, compound tartrazine solution, chloroform spirit, lemon syrup, purified water, syrup, etc.

Formula

Codeine phosphate	0.3 g
Benzoic acid	2.0 ml
Compound tartrazine solution	1.0 ml
Chloroform spirit	2.0 ml
Lemon syrup	20.0 ml
Purified water	2.0 ml
Syrup, sufficient to produce	100.0 ml

Procedure

1. Dissolve the codeine phosphate in purified water.
2. Add 50 ml of syrup and mix.
3. Add Benzoic acid, compound tartrazine solution, chloroform spirit, and lemon syrup.
4. Add sufficient syrup to produce 100 ml.

Category : Expectorant

Dose : 5 ml. When a dose is prescribed which is less than or not a multiple of 5.0 ml. Codeine linctus may be diluted with syrup.

Storage : Preserve in a well-closed container and protected from light.

Uses : Expectorant

EXERCISE NO. 2

Object : To prepare Pholcodine linctus

Requirements : Pholcodine, citric acid monohydrate, chloroform spirit, amaranth solution, compound tartrazine solution, syrup, etc.

Formula

Pholcodine	0.10 g
Citric acid monohydrate	1.0 g
Chloroform spirit	7.5 ml
Amaranth solution	0.10 ml
Compound tartrazine solution	1.0 ml
Syrup, sufficient to produce	100.0 ml

Procedure

1. Dissolve the pholocodine in 3.5 ml of the chloroform spirit.
2. Separately dissolve the citric acid monohydrate in 4.0 ml of chloroform spirit.
3. Mix these two solutions.
4. Add amaranth solution, and compound tartrazine solution and mix.
5. Add sufficient quantity of syrup to produce 100 ml.

Category : Expectorant

Weight per ml : 1.290 to 1.310 g

Dose : 5 ml. When a dose is prescribed which is less than or not a multiple of 5.0 ml. Codeine linctus may be diluted with syrup. Pholcodeine linctus contains in 5 ml 5 mg of pholcodiene.

Storage : Preserve in a well-closed container and protected from light.

Uses : Expectorant

EXERCISE NO. 3

Object : To prepare Simple linctus

Requirements : Concentrated anise water, amaranth solution, citric acid monohydrate, chloroform spirit, syrup, etc.

Formula

Concentrated anise water	1.0 ml
Amaranth solution	1.5 ml
Citric acid monohydrate	2,5 g
Chloroform spirit	6.0 ml
Syrup, sufficient to produce	100.0 ml

Procedure

1. Dissolve the pholocodine in 3.5 ml of the chloroform spirit.
2. Separately dissolve the citric acid monohydrate in 4.0 ml of chloroform spirit.
3. Mix these two solutions.
4. Add amaranth solution, and compound tartrazine solution and mix.
5. Add sufficient quantity of syrup to produce 100 ml.

Category : Carminative **Weight per ml :** 1.270 to 1.310 g

Dose : 5 ml. When a dose is prescribed which is less than or not a multiple of 5.0 ml. Codeine linctus may be diluted with syrup.

Storage : Simple linctus should be stored at a temperature not exceeding 25°

Uses : Carminative

EXERCISE NO. 4

Object : To prepare Paediatric simple linctus

Requirements : Concentrated anise water, amaranth solution, citric acid monohydrate, chloroform spirit, syrup, etc.

Formula

Simple linctus	25.0 ml
Syrup, sufficient to produce	100.0 ml

Procedure

1. Mix 25.0 ml of simple linctus and syrup.
2. Add sufficient syrup to produce 100 ml.

Category : Carminative

Weight per ml : 1.280 to 1.330 g

Dose : 5 ml. When a dose is prescribed which is less than or not a multiple of 5.0 ml. Codeine linctus may be diluted with syrup.

Storage : Simple linctus should be stored at a temperature not exceeding 25°C.

Uses : Carminative

EXERCISE NO. 5

Object : To prepare Paediatric opiate squill linctus

Requirements : Squill oxymel (Squill, water, acetic acid, and honey), camphorated opium tincture, tolu syrup, glycerol, syrup, etc.

Formula

Squill oxymel	6.0 ml
Camphorated opium tincture	6.0 ml
Tolu syrup	6.0 ml
Glycerol	20.0 ml
Syrup, sufficient to produce	100.0 ml

Procedure

1. Mix squill oxymel with camphorated opium tincture.
2. Add gradually tolu syrup and glycerol.
3. Add syrup sufficient to produce 100 ml and mix.

Category : Seizure and anemia **Weight per ml :** 1.270 to 1.300 g

Dose : For child 5 to 10 ml. When a dose is prescribed which is less than or not a multiple of 5.0 ml. Codeine linctus may be diluted with syrup. Paediatric opiate squill linctus contains, in 5 ml. 150 μg of anhydrous morphine

Storage : Simple linctus should be stored at a temperature not exceeding 25°C.

Uses : Seizure disorder and anemia

EXERCISE NO. 6

Object : To prepare Paediatric Compound Tolu linctus

Requirements : Benzaldehyde spirit, citric acid monohydrate, compound tartrazine solution, glycerol, Invert syrup, Tolu syrup, etc.

Formula

Benzaldehyde spirit	0.1 ml
Citric acid monohydrate	0.4 g
Compound tartrazine solution,	1.0 ml
Glycerol	20.0 ml
Invert syrup	20.0 ml
Tolu syrup, sufficient to produce	100.0 ml

Procedure

1. Mix all ingredients except Tolu syrup gradually one by one in a container.
2. Add Tolu syrup sufficient to produce 100 ml.

Category : Carminative **Weight per ml :** 1.300 to 1.320 g

Dose : For child 5 to 10 ml. When a dose is prescribed which is less than or not a multiple of 5.0 ml. Codeine linctus may be diluted with syrup.

Storage : Simple linctus should be stored at a temperature not exceeding 25°C.

Uses : Carminative

EXERCISE NO. 7

Object : To prepare Opiate squill linctus

Requirements : Squill oxymel, camphorated opium tincture, Tolu syrup, etc.

Formula

Squill oxymel	30.0 ml
Camphorated opium tincture	30.0 ml
Tolu syrup	30.0 ml

Procedure

1. Mix ingredients alternatively in a flask.
2. Disperse 0.1 g of xanthum gum or 0.5 g of powdered tragacanth in the camphorated opium tincture.
3. Add the squill oxymel and Tolu syrup, add and mix properly.

Category : Carminative

Dose : 5 ml. When a dose is prescribed which is less than or not a multiple of 5.0 ml. Codeine linctus may be diluted with syrup.

Storage : Preserve in a well-closed container.

Uses : Carminative

EXERCISE NO. 8

Object : To prepare Methadone linctus

Requirements : Methadone hydrochloride, compound tartrazine solution, glycerol, purified water, Tolu syrup, etc.

Formula

Methadone hydrochloride	40.0 mg
Compound tartrazine solution,	0.80 ml
Glycerol	25.0 ml
Purified water	12.0 ml
Tolu syrup, sufficient to produce	100.0 ml

Procedure

1. Dissolve methadone hydrochloride in purified water.
2. Add the compound tartrazine solution.
3. Add glycerin and Tolu syrup sufficient to produce 100 ml.

Category : Analgesic

Weight per ml : 1.260 to 1.280 g

Dose : Methadone linctus contains in 5 ml 2.0 mg of Methadone hydrochloride.

Storage : Preserve in a well-closed container.

Uses : Analgesic and antitussive

EXERCISE NO. 9

Object : To prepare Noscapine linctus

Requirements : Noscapine, citric acid monohydrate, compound tartrazine solution, chloroform spirit, purified water, syrup, etc.

Formula

Noscapine	0.30 g
Citric acid monohydrate	1.00 g
Compound tartrazine solution	0.30 ml
Chloroform spirit	7.50 ml
Purified water	10.0 ml
Syrup, sufficient to produce	100.0 ml

Procedure

1. Mix noscapine, citric acid monohydrate to the priority, heated water purified water at 50°C and stirrer until dissolved.
2. Cool and add 80 ml of syrup and mix.
3. Add compound tartrazine solution and chloroform spirit.
4. Add syrup sufficient to produce 100 ml.

Category : Antitussive

Dose : For child 5 to 10 ml. When a dose is prescribed which is less than or not a multiple of 5.0 ml, Codeine linctus may be diluted with syrup.

Storage : Simple linctus should be stored at a temperature not exceeding 25°C.

Uses : As antitussive, it is used in the management of cough in bronchial asthma and pulmonary emphysema.

VIVA VOCE QUESTIONS

For Exercise No. 1 to 9.

1. What are linctuses?
2. How linctuses are prepared?
3. What is the general application of the linctus?
4. What is the use of simple linctus?
5. Why codeine linctus is required?
6. What is the difference between linctus and syrup?
7. What is the use of Tolu syrup?
8. What is antitussive?
9. What is the diluent used in linctus?
10. Which type of container is suitable for linctus?
11. What is Noscapine?
12. What is preservative?
13. What is the use of citric acid monohydrate?
14. What is Invert syrup?
15. What is the composition of squill oxymel?

9

Liniments

- White liniment
- Camphor liniment
- Soap liniment
- Turpentine liniment
- Ammonium camphor liniment
- Methyl salicylate liniment
- Aconite liniment
- Belladona liniment
- Calamine liniment

Liniments are solution or mixture of various substances in oil, alcoholic solution of soap or emulsions or occasionally semi-solid preparations intended for external application and should be labeled accordingly. They are applied with rubbing or massaged into the skin as counter irritating or stimulating agents to the effected area and because of this were known as embrocations. Some liniments are applied on a warm dressing or with a brush like analyzer and soothing type.

Dental liniments, which are no longer official are solutions of active substances and are applied onto the gums by rubbing. Liniments are usually applied with fraction and rubbing of the skin, the oil or soap base proving for easy of application and massage. Alcoholic liniments are used generally for their rubifacient, counter irritant, mildly astringent and shows penetrating effects. These types of liniments easily penetrate to the skin than those with oil base.

The oily liniments therefore, are slow in their action but are more useful when massaged. The function of liniment depends on the additives but most of liniments may function solely as protective coating on the affected area. Liniment should not be applied to the broken or bruised skin because they would be very irritant especially if the solvent used as alcohol.

They may contain following substances such as

(a) Analgesic (b) Antimicrobial (c) Rubefacient
(d) Counter irritant (e) Stimulants (f) Soothing agents

Although alcohol is primarily used as solvent, it enhances the penetration of the medicaments into the skin and has counter irritants or rubificient action. Counter irritants are used to mask pain from fibrositis, sciatica, nuralgia and similar complaints by producing warmth, tingling and numbness. When rubbed onto the skin, they also cause redness, are called as rubifacients. Cottonseed oil and arachis oil are less irritant than alcohol and spread more easily on the skin.
Two types of vehicle are used for the preparation of liniments.

1. Alcohol e.g. soap liniment and aconite liniment
2. Oils e.g. camphor liniment and methylsalicyalte

LABELLING

It should comply with the general requirements for labelling. In addition the label on the container should indicates

FOR EXTERNAL USE ONLY

SHAKE WELL BEFORE USE ONLY

NOT TO BE APPLIED TO WOUNDS OR BROKEN SKIN

STORE IN COOL PLACE

INFLAMMABLE

CONTAINER

Narrow mouthed screw capped colored bottles can be used for storing liniments.

EXERCISE NO. 1

Object : To prepare White liniment

Requirements : Oleic acid, turpentine oil, ammonia solution, ammonium chloride and common glasswares

Formula

Oleic acid	85.0 ml
Turpentine oil	250.0 ml
Dilute ammonia solution	45.0 ml
Ammonium chloride	12.5 g
Purified water	625.0 ml

Procedure

1. Mix the oleic acid with the measured quantity of turpentine oil.
2. Mix dilute ammonia solution with 45 ml of purified water and warm it.
3. Add warm diluted ammonia solution to the oily solution and shake to form an emulsion.
4. Dissolve ammonium chloride in remaining amount of purified water.
5. Mix this solution in the emulsion and shake it.

Category : Counter irritants

Storage : Certain plastic containers are not suitable for white liniments, use narrow mouthed colored glass bottle.

Uses : Counter irritants

VIVA VOCE QUESTIONS

1. What is liniment?
2. Give pharmaceutical application of liniment.
3. How will you prepare liniment in the laboratory?
4. Give storage condition of the liniment.
5. Give application of various additives in the white liniment.
6. What is counter irritant?
7. Define rubifacient with examples.
8. What is the use of oleic acid in the preparation of liniment?

EXERCISE NO. 2

Object : To prepare Camphor liniment

Synonyms : Camphorated oil

Requirements : Camphor, arachis oil and common glasswares

Theory : Camphor is a ketone, obtained from *cinnamomum camphora* Linn. It contains not less than 96.0 percent of $C_{10}H_{16}O$. It is a colorless or white crystals, granules, or crystalline masses. It is volatile in nature therefore, the preparation of camphor liniment should be stored in a well closed container at a cool place. It has a penetrating characteristic odor, a pungent, aromatic taste and is readily pulverizable in the presence of small amount of alcohol and other organic solvents. It is used locally as antiseptic, mildly anaesthetic, and rubefacient when rubbed on the skin. It is also used as a counter irritant for inflamed joint, sprains, rheumatic and other inflammatory conditions such as colds in the throat and chest. Arachis oil is obtained from the seed kernels of one or more of the cultivated varieties of *arachis hypogaea* Linn. It is colorless or pale yellow oily liquid with a characteristic of nutty odor and has a bland taste. It is used as a solvent for the preparation of various pharmaceutical products. It is used for the preparation of liniments, plasters, ointments and soaps.

Formula

Camphor	20.0 g
Arachis oil	80.0 g

Procedure

Mix weighed amount of camphor in arachis oil in a closed vessel

Category : Counter irritants

Storage : Preserve it in a well-closed glass container and store in a cool place

Uses : Counter irritants

VIVA VOCE QUESTIONS

1. What is the use of camphor liniment?
2. Why arachis oil is used in this liniment?
3. What is the difference between arachis oil and alcohol as a solvent?
4. Why liniment is to be stored in closed container?
5. What do you mean by counter irritant?

EXERCISE NO. 3

Object : To prepare Soap liniment

Requirements : Soft soap, camphor, lemon grass oil, alcohol and common glasswares

Formula

Soft soap	8.0 g
Camphor	4.0 g
Lemon grass oil	1.5 ml
Purified water	17.0 ml
Alcohol (90.0%), sufficient to produce	100.0 ml

Procedure

1. Dissolve the weighed quantity of soft soap, camphor and lemon grass oil in 60 ml of alcohol.
2. Add purified water and remaining amount of alcohol to make up the volume and mix.
3. Keep aside for a week and filter to remove the undissolved substance.

Category : Detergent, Mild local irritant

Storage : Preservative in a well-closed container.

Uses : Mild local irritant

VIVA VOCE QUESTIONS

1. Why soap liniment are prepared?
2. How to prepare soap liniment?
3. What is the use of soft soap in this preparation?
4. What is the use of camphor in this preparation?
5. Which type of container is more suitable for these preparation?
6. What is the requirements of label for liniment
7. How it differs from lotions?
8. What is the use of lemongrass oil?

EXERCISE NO. 4

Object : To prepare Turpentine liniment

Synonym : Turpent Liniment

Requirements : Soft soap, Camphor, turpentine oil, etc.

Theory : Turpentine oil is the volatile oil distilled from the Oleo-resin obtained from *Pinus palustries* Miller. and also obtained from other species of pinus. It is colorless, limpid liquid with characteristic odor, pungent taste and some what bitter.

On exposure to air and moisture it under goes rapid changes in viscosity and convert in a yellow viscous oil and also changes due to the chemical reaction. This is the main reason to use freshly prepared rectified oil in the liniment.

It is used as a counter irritant. In this preparation, soft soap is used as emulsifying agent. It is a monovalent soap and forms o/w type of emulsion. Soaps are salts of fatty acids. Counter irritant are those substances which act against irritation and removes pain. Rubifacient are those substances which redden the skin.

Formula

Soft soap	9.0 g
Camphor	5.0 g
Turpentine oil	65.0 ml
Purified water, sufficient to produce	100.0 ml

Procedure

1. Mix the soft soap with small amount of purified water (10 ml).
2. Make solution of camphor in fresh rectified turpentine oil.
3. Gradually add camphor solution to the soap mixture with trituration until a thick creamy emulsion is formed.
4. Add sufficient amount of purified water to make up the volume and mix.

Category : Counter irritant

Storage : Preserve in a well-closed container and protected from light and store in col place.

Uses : Counter irritant and rubefacient.

VIVA VOCE QUESTIONS

1. What is turpentine oil?
2. What is emulsifying agent?
3. What is the role of soap in this preparation?
4. Give pharmaceutical application of liniment.
5. Define counter-irritant and rubefacient.
6. Why the liniment is stored in well-closed container?
7. How liniment is prepared?
8. Give application of alcohol in this preparation.
9. What is the dual action of alcohol?

EXERCISE NO. 5

Object : To prepare Ammoniated camphor liniment

Requirements : Camphor, eucalyptus oil, alcohol (90.0%), ammoniated strong solution

Theory : Eucalyptus oil is the oil distilled from the fresh leaves of *Eucalyptus globulus* Labill. Or from the other species of *Eucalyptus* and rectified. It contains not less than 60.0 percent w/w of cineole $C_{10}H_{18}O$. It is a colorless or pale yellow liquid, with aromatic and camphoraceous odor. Taste is pungent and camphoraceous, followed by a sensation of cold. In this preparation strong solution of ammonia acts as counter-irritant and rubifacient. Camphor acts as counter-irritant and eucalyptus oil acts as flavoring agent and also rubifacient. Alcohol is used as vehicle and it enhances the action as well as penetration power of the preparation. This preparation should be stored in a well-closed container because it contains volatile ingredients.

Formula

Camphor	12.5 g
Eucalyptus oil	0.50 ml
Ammonia solution strong	25.0 ml
Alcohol (90%), sufficient to produce	100.0 ml

Procedure

1. Dissolve the camphor and the eucalyptus oil in 60.0 ml of alcohol (90%).
2. Add the measured quantity of the strong ammonia solution with frequent shaking.
3. Make up the volume by adding alcohol and mix it.

Category : Counter irritant

Storage : Preserve ammoniated camphor liniment in a well-closed glass stoppered bottle.

Uses : Counter irritant

VIVA VOCE QUESTIONS

1. Why liniment can not be applied on broken skin?
2. Which vehicle is suitable in place of alcohol?
3. What is the source of eucalyptus oil?
4. How liniment are applied?
5. Give application of liniment.
6. How liniments are stored?

EXERCISE NO. 6

Object : To prepare Aconite liniment

Requirements : aconite, camphor and alcohol (90.0%) and common glasswares

Theory : Aconite is obtained from the dried root of *Aconitum napellus* Linn. It has slight odor and produces a persistent tingling sensation followed by numbless. Aconite and its preparation are poisonous and rarely used internally except in homeopathy.

Aconite is used externally in neuralgia, rheumatism and inflammation in the form of liniment of aconite. Tincture aconite is used as an antipyretic in small doses. Aconite is depressant to the heart. In higher doses it causes at first burning and tingling sensation on the entire body which becomes intolerable, which followed by anesthetic action on the body and a feeling of cold as if ice water circulates instead of blood. Pulse becomes slow and death occurs due to respiratory failure.

Formula

Aconite, in moderately coarse powder	50.0 g
Camphor	3.0 g
Alcohol (90.0%), sufficient to produce	100.0 ml

Procedure

1. Exhaust the aconite powder with alcohol by percolation and reserve the first 75.0 ml of the percolate.
2. Evaporate the remainder quantity under reduced pressure to a syrupy consistence.
3. Add the reserve portion and mix.
4. Dissolve the weighed camphor in the mixture.
5. Add sufficient quantity of alcohol to produce the required volume and set aside for not less than twenty four hours.
6. Filter the preparation and fill in a suitable container with label.

Category : Counter irritant and analgesic

Weight per ml. : At 20° C, 0.870 to 0.900 g.

Storage : Preserve aconite liniment in a well closed container and store in a cool place.

Uses : Counter irritant and analgesic

VIVA VOCE QUESTIONS

1. What is aconite?
2. How aconite liniment is prepared?
3. Why aconite liniment is prepared?
4. Give application of aconite.
5. What is the role of alcohol in this liniment?
6. Which type of vehicle is produces less irritation as compare to alcohol?

Note : In the preparation of this liniment the alcohol (90%) may be replaced by specially denatured spirit diluted so as to be of equivalent alcoholic strength.

EXERCISE NO. 7

Object : To prepare Belladonna liniment

Requirements : Belladonna liquid extract, camphor, and alcohol (80.0%) and common glasswares.

Theory : Belladonna liquid extract contains 0.75 percent w/v of the alkaloids of belladonna root, calculated as hyoscyamine. It is obtained by the percolate of belladonna root with alcohol (80.0%). Hyoscyamine has local anaesthetic and anodyne action and so liniment and plaster of belladonna used to remove peripheral pain.

Formula

Belladonna liquid extract	50.0 ml
Camphor	5.0 g
Alcohol (80.0%)	50.0 ml

Procedure

1. Dissolve belladonna liquid extract in alcohol.
2. Add camphor in the mixture and mix it.
3. Add remaining quantity of alcohol to make up the volume.

Category : local anaesthetic and parasympatholytic

Storage : Preserve belladonna liniment in a well-closed container and protected from light.

Uses : local anaesthetic and anodyne action to remove peripheral pain

VIVA VOCE QUESTIONS

1. What is belladonna?
2. How will you prepare liniment of belladonna?
3. What is the pharmacology of belladonna?
4. What is the use of belladonna liniment?
5. Give pharmaceutical application of belladonna.
6. How will you store belladonna liniment?
7. What is anodyne?
8. What is local anaesthetic agent?

EXERCISE NO. 8

Object : To prepare Calamine liniment

Requirements : Calamine, wool fat, oleic acid, arachis oil and calcium hydroxide solution.

Formula

Calamine	5.0 g
Wool fat	1.0 g
Oleic acid	0.5 ml
Arachis oil	50.0 ml
Calsium hydroxide solution	100.0 ml

Procedure

1. Melt weighed amount of wool fat, oleic acid and arachis oil.
2. Triturate calamine with the melted oil.
3. Add calcium hydroxide solution and shake it.
4. Transfer it to a suitable container and shake vigorously.

Category : Rubefacients

Storage : Preserve calamine liniment in a well-closed container and protected from light.

Uses : It is prescribed by the dermatologists as rubefacients and also used by consumers for treatment of minor muscular aches and pains.

VIVA VOCE QUESTIONS

1. What is calamine?
2. How calamine liniment is prepared?
3. What is difference between liniment and lotion?
4. Why in olden days liniment was less popular than lotion?
5. What is rubefacient and muscular aches?
6. What is pain?
7. What is the pharmacology of pain?
8. What is the use of arachis oil?
9. Why calamine liniment is prepared?
10. What is the role of wool fat?

10

Lotions

• Calamine lotion	• Oily calamine lotion
• Boric acid lotion	• Zinc sulphate lotion
• Acriflavine lotion	• Potassium permanganate lotion
• Cetrimide lotion	• Salicylic acid lotion
• Dichloroxylenol lotion	• Thiomersal lotion
• Gentian violet lotion	

Lotions are usually liquid or liquid suspensions or semi-solid preparation containing one or more medicaments, intended to be applied to the uniform skin without friction. They are either dabbed on the skin or applied on a suitable dressing and covered with waterproof substance to reduce the evaporation. They may be prepared by triturating the ingredients to a smooth paste and then gradually adding the remaining liquid phase. For large quantity preparation of lotion, high-speed mixers or homogenizers produce better quality of lotions. The particles in colloidal dimension are more soothing to inflamed areas and are more effective in contact with infected surfaces. A wide variety of ingredients are employed in the preparation to produce better dispersions that shows good cooling, soothing, drying or protective nature of the lotion. Following substances are used in the preparation of lotions.

1. Bentonite : used as a suspending agent.
2. Methylcellulose or sodium carboxymethylcellulose : used to hold the active ingredient in contact with the affected site.
3. Glycerin : keep the skin moist for considerable period of time.
4. Alcohol : used to accentuated action like drying, cooling etc.
5. Miscellaneous : benzocaine, calamine, resorcin, steroids, sulphur, zinc oxide, etc.

Lotions are generally prescribed for the following purpose.

1. Anesthetic
2. Antiseptics
3. Astringents
4. Germicides
5. Protective
6. Antihistaminic
7. Screening agents

Microorganism may grow in certain lotions if no preservative is included in the preparation. Care should be taken to avoid contamination during the preparation of lotion, even it contains preservative.

Dilutions of Lotions : Care should be taken in dilutions, particularly to prevent microbial contamination. The appropriate diluent should be used and heating should be avoided during mixing. Diluted lotions should be used within four weeks or their preparations.

Examples

Acriflavine lotion	Dichloroxylenol lotion	Salicylic acid lotion
Boric acid lotion	Gentian violet lotion	Thiomersal lotion
Calamine lotion	Oily calamine lotion	Zinc sulphate lotion
Cetrimide lotion	Phenolated lotion	White lotion
Dichloroxylenol lotion	Potassium permanganate lotion	

LABELLING

Comply with the general requirements for labelling. In addition the label on the container states.

1. 'Shake Well Before Use'
2. 'For External Use Only'

EXERCISE NO. 1

Object : To prepare Calamine lotion

Requirements : Calamine, zinc oxide, bentonite, sodium citrate, liquefied phenol, glycerin, rose water, etc.

Theory : Calamine is a basic zinc carbonate suitably colored with ferric oxide. It is used as mild astringent. Bentonite is a colloidal hydrated aluminium silicate and acts as a suspending agent. Zinc oxide acts as sedative and productive of skin. Phenol to used as antiseptic and preservatives.

Glycerin has soothing effect. Sodium citrate is used as chelating agent to chelate the Iron, as free iron can change the color of the preparation by reacting with phenol. The preparation is stored in a well closed container because Zinc oxide gradually converts into zinc carbonate due to atmospheric carbon dioxide. Rose water is used as vehicle and flavoring agent.

Formula

Calamine	15.0 g
Zinc oxide	5.0 g
Bentonite	3.0 g
Sodium citrate	0.5 ml
Liquefied phenol	0.5 ml
Glycerin	5.0 ml
Rose water, sufficient to produce	100.0 ml

Procedure

1. Prepare sodium citrate solution in 70 ml rose water.
2. Triturate calamine, zinc oxide, and bentonite with citrate solution.
3. Add liquefied phenol and mix.
4. Add glycerin and sufficient rose water to produce 100 ml.

Category : Protective

Storage : Preserve in a well-closed container.

Uses : Protective

EXERCISE NO. 2

Object : To prepare Boric acid lotion

Requirements : Chlorinated lime, boric acid, purified water, etc

Formula

Chlorinated lime	1.25 g
Boric acid	1.25 g
Purified water, sufficient to produce	100.0 ml

Procedure

1. Mix chlorinated lime and boric acid.
2. Dissolve in purified water and add sufficient water to produce 100 ml.

Category : Local anti-infective, germicide

Storage : Preserve in a well-closed container.

Uses : Local anti-infective, germicide

EXERCISE NO. 3

Object : To prepare Acriflavine lotion

Requirements : Acriflavine, industrial methylated spirit, purified water, etc.

Formula

Acriflavine	200 mg
Industrial methylated spirit	75.0 ml
Purified water, sufficient to produce	100.0 ml

Procedure

1. Mix acriflavine with industrial methylated spirit.
2. Add purified water sufficient to produce 100 ml.

Category : Protective **Storage :** Preserve in a well-closed container.

Uses : Protective

EXERCISE NO. 4

Object : To prepare Cetrimide lotion

Requirements : Cetrimide, purified water, etc.

Formula

Cetrimide	1.0 mg
Purified water, sufficient to produce	100.0 ml

Procedure

1. Dissolve cetrimide in purified water and make up the volume using purified water.

Category : Antibacterial

Storage : Preserve in a well-closed container.

Uses : Antibacterial

EXERCISE NO. 5

Object : To prepare Dichloroxylenol lotion

Requirements : Dichloroxylenol, terpinol, ethanol, purified water, etc.

Formula

Dichloroxylenol	5.0 g
Terpinol	10.4 ml
Ethanol	20.0 ml
Purified water, sufficient to produce	100.0 ml

Procedure

1. Dissolve dichloroxylenol in ethanol (95%).
2. Add terpinol and mix.
3. Add sufficient purified water to produce 100 ml.

Category : topical antiseptics **Storage :** Preserve in a well-closed container.

Uses : topical antiseptics

EXERCISE NO. 6

Object : To prepare Gentian violet lotion

Requirements : Gentian violet, ethanol, purified water, etc.

Formula

Gentian violet	1.0 g
Ethanol (95%)	10.0 ml
Purified water, sufficient to produce	100.0 ml

Procedure

1. Dissolve gentian violet in ethanol (95%).
2. Add sufficient purified water to produce 100 ml.

Category : Pharmaceutical aid

Storage : Preserve in a well-closed container.

Uses : Pharmaceutical aid

EXERCISE NO. 7

Object : To prepare Oily calamine lotion

Requirements : Calamine, oleic acid, wool fat, arachis oil, calcium hydroxide solution, etc.

Formula

Calamine	5.0 g
Oleic acid	0.5 ml
Wool fat	1.0 g
Arachis oil	50.0 ml
Calcium hydroxide solution, sufficient to produce	100.0 ml

Procedure

1. Melt oleic acid, wool fat and arachis oil in a container
2. Triturate the calamine with the mixture.
3. Transfer to a suitable container.
4. Add calcium hydroxide solution sufficient to produce 100 ml and shake vigorously.

Category : Protective

Storage : Preserve in a well-closed container.

Uses : Protective

EXERCISE NO. 8

Object : To prepare Zinc sulphate lotion

Requirements : Zinc sulphate, amaranth solution, purified water, etc.

Formula

Zinc sulphate	1.0 g
Amaranth solution	1.0 ml
Purified water, sufficient to produce	100.0 ml

Procedure

1. Dissolve zinc sulphate in purified water.
2. Add amaranth solution and mix.
3. Add sufficient purified water to produced 100 ml.

Category : Protective

Storage : Preserve in a well-closed container.

Uses : Protective

EXERCISE NO. 9

Object : To prepare Potassium permanganate lotion

Requirements : Potassium permanganate,

Formula

Potassium permanganate	0.1 g
Purified water, sufficient to produce	100.0 ml

Procedure

1. Triturate potassium permanganate with purified water.
2. Add sufficient purified water to produce 100 ml.

Category : Antiseptic

Storage : Preserve in a well-closed container. It must be freshly prepared. Diluted with 7 parts of water or as directed.

Uses : Antiseptic

EXERCISE NO. 10

Object : To prepare Salicylic acid lotion

Requirements : Salicylic acid, castor oil, ethanol, etc.

Formula

Salicylic acid	2.0 g
Castor oil	1.0 ml
Ethanol (96%), sufficient to produce	100.0 ml

Procedure

1. Dissolve the salicylic acid in a portion of ethanol (96%).
2. Add the castor oil.
3. Add ethanol (96%) sufficient to produce 100 ml and mix.

Category : Analgesic
Storage : Preserve in a well-closed container. The label on the container or the label on the package states that the preparations is flammable and should not be used, nor the hair dried, near a fire or flamed flame.

Uses : Analgesic

EXERCISE NO. 11

Object : To prepare Thiomersal lotion

Requirements : Thiomersal, ethanol, etc.

Formula

Thiomersal	0.1 g
Ethanol (90%), sufficient to produce	100.0 ml

Procedure

1. Triturate thiomersal with ethanol.
2. Add sufficient ethanol to produce 100 ml.

Category : Antibacterial agent

Storage : It should be dispensed in tightly closed containers. **Uses :** Antibacterial

VIVA VOCE QUESTIONS

For Exercise No. 14 to 24.

1. What is calamine lotion?
2. What is the pharmaceutical application of calamine?
3. What is the use of bentonite and zinc oxide in lotion?
4. What is the use of boric acid lotion?
5. What is difference between lotion and liniment?
6. What is acriflavine lotion.
7. How applied lotion?
8. Why lotion can not be applied on the broken skin?
9. Why cetrimide lotion is prepared?
10. What is the use of gentian violet?
11. Which solvent is most suitable in place of alcohol in lotion?
12. How to apply lotion on the skin?
13. What is astringent?

11

Ointments

- Calamine ointment
- Calamine and coal tar ointment
- Ichthammol ointment
- Cetrimide emuslifying ointment
- Emulsifying ointment
- Simple ointment
- Paraffin ointment
- Wool alocihol ointment
- Salicylic acid ointment
- Sulphur ointment
- Zinc and castor oil ointment
- Coal tar and salicylic acid ointment
- Coal tar and Zinc oxide ointment

Ointments are semi-solid preparation for application to the skin or mucous membranes. They are often anhydrous and contain the medicament either dissolved or dispersed in the vehicle. The solid substances, which are dispersed, should be in the form of a micronized powder. They are easily spread and their plastic viscosity may be controlled by modifications of the formulation. Ointments are typically used as emollient, protective and vehicles to incorporate medicaments.

OINTMENT BASES

Ointment bases are mainly four types

1. **Hydrocarbon bases or oleaginous**

   They are emollient, occlusive, nonwater-washable, hydrophobic and greasy.
   e.g. White petrolium, glyceryl monostearate, isopropyl myristate, isopropyl palmitate, butyl stearate, cetyl alcohol, stearyl alcohol, polyethylene glycol, lanolin, and hydrogenated lanolin.

2. **Absorption bases**
   They are emollient, water absorbent, occlusive, anhydrous, and greasy.
   e.g. hydrophilic petrolatum (Aquaphor, polysorb, kessolin, etc.), wool fat, etc.

3. Emulsifying bases

a. W/O emulsions : They are emollient, occlusive, contain water, some absorb additional water and greasy. e.g. hydrous wool fat, cold cream, etc.
b. O/W emulsions : They are water-washable, non-greasy, can diluted with water, and nonocclusive. e.g. Hydrophilic ointment, vanishing cream, etc.

4. Water soluble bases

They are usually anhydrous, water-soluble and washable, non-greasy, non-occlusive and lipid free. e.g. polyethylene glycol, propylene glycol, propylene glycol-ethanol, etc.

METHOD OF PREPARATIONS

1. **Trituration** : When the quantity of ointment is not more than 50 g, a white porcelain or marble ointment should be used in conjunction with a flexible steel spatula. A steel spatula should not be used as medicament may react with the metal. Following substances react with metal such as mercury compounds, tannic acid, salicylic acid and iodine.

2. **Fusion :** Ointment containing hard paraffin, beeswax, emulsified wax, wool alcohol are prepared by melting ingredients in a porcelain dish over a water bath. In this process higher melting point substance should be melted first and add then other ingredients of the bases in order of their melting point.

PACKAGING AND LABELLING

1. Ointment jars

a. Straight-sided screw cap jars of glass or plastic.
b. Clear, amber or opaque glasses containers
c. White, opaque, plastic usually high-density polyethylene jar.
d. Metal or composition plastic tops

2. Ointment tubes

a. Made of tin or aluminium of plastic materials e.g. polyethylene, polypropylene c.
b. Ointment tube should be packed in a winged plastic board box..
c. The box serves to hold and protect the ointment tube and to bear carry the label.
d. The ointment tube is identified with an inked-on prescription number so that both tube and box are identified.

On large scale tubes are labeled in a variety of ways. Paper labels may be used. Label may be silk-screened onto plastic surfaces. Expiry dates and code lot numbers may be stamped on as a part of the tube crimping procedure.

EXERCISE NO. 1

Object : To prepare Calamine ointment

Requirements : calamine, white soft paraffin, etc

Formula

Calamine, finely sifted	1.50 g
White soft paraffin	8.50 g

Procedure

1. Triturate the calamine powder with part of the white soft paraffin until smooth.
2. Add gradually sufficient quantity of white soft paraffin to produce 10 g.

Category : Astringent and protective

Storage : Preserve in a well-closed container.

Uses : Astringent and protective

EXERCISE NO. 2

Object : To prepare Calamine and coal tar ointment

Requirements : Calamine, zinc oxide, strong coal tar solution, hydrous wool fat, white soft paraffin, etc.

Formula

Calamine, finely sifted	1.25 g
Zinc oxide, finely sifted	1.25 g
Strong coal tar solution	0.25 g
Hydrous wool fat	2.50 g
White soft paraffin	4.75 g

Procedure

1. Melt hydrous wool fat and white soft paraffin in a glass container (100 ml beaker).
2. Powdered calamine and zinc oxide incorporate in the melted base.
3. Add coal tar solution and stir gently until cold.

Category : Astringent and protective

Storage : Preserve in a well-closed container and minimize evaporation losses during the storage.

Uses : Astringent and protective

EXERCISE NO. 3

Object : To prepare Ichthammol ointment

Requirements : Ichthammol, wool fat, yellow soft paraffin, etc.

Formula

Ichthammol	1.00 g
Wool fat	4.50 g
Yellow soft paraffin	4.50 g

Procedure

1. Melt together wool fat and yellow soft paraffin in a glass container.
2. Add Ichthammol and stir until cold.

Category : Astringent and local antibacterial

Storage : Preserve in a well-closed container and store in cool place.

Uses : It is mildly astringent, irritant, and local antibacterial agent with moderately emollient and demulcent properties.

EXERCISE NO. 4

Object : To prepare Cetrimide emulsifying ointment

Requirements : White soft paraffin, cetostearyl alcohol, liquid paraffin, cetrimide, etc

Formula

Cetrimide	0.30 g
White soft paraffin	5.00 g
Cetostearyl alcohol	2.70 g
Liquid paraffin	2.00 g

Procedure

1. Melt together white soft paraffin, cetostearyl alcohol, and liquid paraffin.
2. Add powdered cetrimide in base and stir until cold.

Category : Protective

Storage : Preserve in a well-closed container and store in cool place.

Uses : Protective

EXERCISE NO. 6

Object : To prepare Emulsifying ointment

Requirements : Liquid paraffin, emulsifying wax, white soft paraffin, etc.

Formula

Liquid paraffin,	2.00 g
Emulsifying wax	3.00 g
White soft paraffin	5.00 g

Procedure

1. Melt together liquid paraffin, emulsifying wax, and white soft paraffin.
2. Stir until cool.

Category : Pharmaceutical aid.

Storage : Preserve in a well-closed container, protect from light.

Uses : Pharmaceutical aid.

EXERCISE NO. 7

Object : To prepare Simple ointment

Requirements : Wool fat, cetostearyl alcohol, hard paraffin, white soft paraffin or yellow soft paraffin, etc.

Formula

Wool fat	0.50 g
Cetostearyl alcohol	0.50 g
Hard paraffin	0.50 g
White soft paraffin or yellow soft paraffin	8.50 g

Procedure

1. Mix wool fat, cetostearyl alcohol, hard paraffin, and white soft paraffin or yellow soft paraffin in a container.
2. Heat the mixture gently with stirring until homogeneous.
3. Stir continually until cold.

Category : Pharmaceutical aid.

Storage : Preserve in a well-closed container, protect from light.

Uses : Pharmaceutical aid.

EXERCISE NO. 8

Object : To prepare Paraffin ointment

Requirements : White beeswax, hard paraffin, cetostearyl alcohol, white soft paraffin.

Formula

White beeswax	0.20 g
Hard paraffin	0.30 g
Cetostearyl alcohol	0.50 g
White soft paraffin	9.00 g

Procedure

1. Mix white beeswax, hard paraffin, cetostearyl alcohol, and white soft paraffin.
2. Heat gently with stirring until homogeneous and stir until cold.

Category : Pharmaceutical aid

Storage : Preserve in a well-closed container, protect from light.

Uses : Pharmaceutical aid and protective.

EXERCISE NO. 9

Object : To prepare Wool alcohol ointment

Requirements : Wool alcohol, white soft paraffin or yellow soft paraffin, hard paraffin, liquid paraffin, etc.

Formula

Wool alcohol	0.60 g
White soft paraffin or yellow soft paraffin	1.00 g
Hard paraffin	2.40 g
Liquid paraffin	6.00 g

Procedure

1. Melt together wool alcohol, white soft paraffin or yellow soft paraffin, hard paraffin, and liquid paraffin.
2. Heat gentle and stir until cold.

Category : Pharmaceutical aid

Storage : Preserve in a well-closed container, protected from light.

Uses : Pharmaceutical aid, protective

EXERCISE NO. 10

Object : To prepare Salicylic acid ointment

Requirements : Salicylic acid, wool alcohol ointment, etc.

Formula

Salicylic acid, finely sifted	0.20 g
Wool alcohol ointment	9.80 g

Procedure

1. Melt the wool alcohol ointment.
2. Add gradually the salicylic acid and stir until cold.

Category : Local anti-infective

Storage : Preserve in a well-closed container, protect from light.

Uses : Local anti-infective

Note : The proportion of hard paraffin, soft paraffin, and liquid paraffin may be varied to produce wool alcohol ointment having suitable properties.
If white ointments is required, white soft paraffin may be used else yellow soft paraffin may be added.

EXERCISE NO. 11

Object : To prepare Sulphur ointment

Requirements : Precipitated sulphur, simple ointment, etc

Formula

Precipitated sulphur	1.00 g
Simple ointment (prepared with white soft paraffin)	9.00 g

Procedure

1. Triturate the precipitated sulphur with a portion of the simple ointment until smooth.
2. Add gradually the remainder of the simple ointment and mix thoroughly.

Category : Scabicide

Storage : Preserve in a well-closed container, protect from light.

Uses : Scabicide

EXERCISE NO. 12

Object : To prepare Zinc ointment

Requirements : Zinc oxide, simple ointment, etc

Formula

Zinc oxide, finely sifted	1.50 g
Simple ointment	8.50 g

Procedure

1. Triturate the zinc oxide with a portion of the simple ointment until smooth.
2. Gradually add simple ointment sufficient to produce 10 g.

Category : Astringent, protective and antiseptic

Storage : Preserve in a well-closed container, protect from light.

Uses : Astringent, protective and antiseptic

EXERCISE NO. 13

Object : To prepare Zinc and castor oil ointment

Requirements : Zinc oxide, castor oil, white beeswax, cetostearyl alcohol, arachis oil.

Formula

Zinc oxide, finely sifted	0.75 g
Castor oil	5.00 g
White beeswax	1.00 g
Cetostearyl alcohol	0.20 g
Arachis oil	3.05 g

Procedure

1. Melt white beeswax, cetostearyl alcohol, and arachis oil.
2. Separately triturate the zinc oxide with a portion of the castor oil until smooth.
3. Add the mixture in melted base and stir until the temperature is about 40°C.

Category : Astringent, protective and antiseptic

Storage : Preserve in a well-closed container, protected from light.

Uses : Astringent, protective and antiseptic.

EXERCISE NO. 14

Object : To prepare Coal tar and salicylic acid ointment

Requirements : Coal tar, polysorbate 80, salicylic acid, emulsifying wax, white soft paraffin, coconut oil and liquid paraffin, etc.

Formula

Coal tar	0.20 g
Polysorbate 80	0.40 g
Salicylic acid	0.20g
Emulsifying wax	1.14 g
White soft paraffin	1.90 g
Coconut oil	5.40 g
Liquid paraffin	0.76 g

Procedure

1. Melt the white paraffin and coconut oil.
2. Add liquid paraffin previously warmed at the same temperature and mix (A).
3. Disperse the coal tar in the polysorbate 80.
4. Incorporate the salicylic acid and mix with the previously melted emulsifying wax (B).
5. Add (A) with stirring to the (B) mix throughly and stir until cold.

Category : Local irritant

Storage : Preserve in a well closed container, protect from light.

Uses : Local irritant, disinfectant

EXERCISE NO. 15

Object : To prepare Coal tar and zinc oxide ointment

Requirements : Strong coal tar solution, zinc oxide, yellow soft paraffin, etc.

Formula

Strong coal tar solution	1.00 g
Zinc oxide, finely sifted	3.00 g
Yellow soft paraffin	6.00 g

Procedure

1. Mix the zinc oxide with the strong coal tar solution.
2. Triturate with a portion of the yellow soft paraffin until smooth.
3. Incorporate the yellow soft paraffin and mix.

Category : Local irritant, protective, antiseptic

Storage : Preserve in a well closed container, protect from light.

Uses : Local irritant, disinfectant, antiseptic and astringent.

VIVA VOCE QUESTIONS

For exercise No.1 to 15

1. What are ointments?
2. How will you prepare ointments?
3. Why ointments are required?
4. What is the use of following substances in the preparation of the ointments such as, wool alcohol, wool fat, hard paraffin, soft paraffin, and liquid paraffin.
5. How ointment base is selected?
6. How will you differentiate between ointment and paste?
7. Give advantages and disadvantages of ointments.
8. Give methods of preparation of the ointment.
9. What is the storage condition ointment?
10. How will you select additives for the preparation of the ointments?
11. What are creams?
12. What are gels?
13. What is the composition of simple ointment?
14. How ointments are sterilized?
15. How will you enhance the consistency of the ointments?

12

Ophthalmic Preparations

- Solution for eye drops
- Zinc sulphate eye drops
- Zinc sulphate and adrenaline eye drops
- Amethocane eye drops
- Atropine sulphate eye drops
- Cocaine hydrochloride eye drops
- Fluorescein eye drops
- Framycetin eye drops
- Homatropine eye drops
- Physostigmine eye drops
- Pilocarpine eye drops
- Silver nitarate eye drops
- Sulphacetamide eye drops
- Prednisolone eye drops
- Sodium chloride eye lotion
- Simple eye ointment
- Chloramphenicol eye ointment

Ophthalmic preparations are sterile products, essentially free from foreign particles, suitably formulated and packaged in suitable container for either topical application to the eyelids or instillation into the cul-de-sac between the eyeball and eyelids. Following ophthalmic preparations are currently used such as solution, suspension, ointment and flat disks.

Following categories of drugs are used in and as ophthalmic preparations.

1. Anti-inflammatory agent
2. Anti-microbial drugs e.g. beta adnergic blocking agent, cycloplegics, miotics, mydriatics and vasoconstructions.
3. Anti-microbial drugs e.g. antibiotics, antiseptics, antivirals, fungicides and sulfonamides
4. Local anaesthetic
5. Others e.g. chelating agents, hyper osmolar agents, diagnostic products, lubricant ointments, artificial tears, eye washes, artificial eye solution and contact lens solution.

EYE DROPS

Eye drops are sterile aqueous or oily solutions or suspension of one or medicaments intended for instillation into the conjuctival sac for diagnostic or therapeutic purposes. Oily eye drops are less used but may nevertheless occasionally be requested. They are usually peppered with sterile castor oil and alkaloids incorporated in a vehicle suitable to be used in the form of the base (e.g. pilocarpine eye drops contains pilocarpine dissolved in sterile castor oil). Certain eye drops may be supplied in a drug, sterile form to reconstituted in an appropriate sterile liquid immediately before use. In such cases the label should state clearly on the container 'Powder for Eye Drops' and should include direction for the use of eye drops.

Aqueous preparation of eye drops contains suitable antimicrobial preservatives at appropriate concentrations because it is intended for use on more than one occasion. The antimicrobial preservatives should be compatible with the other ingredients of the preparations and should remain effective throughout the period of use of the 'Eye Drops'.

FORMULATION

Eye drops may contain the following substances

1. Medicaments : anti-microbial agents, local anaesthetic, diagnostic agents, miotics, midriatic, sulfonamides etc.
2. Vehicle : sterile distilled water, sterile castor oil etc.
3. Viscosity : polyethylene glycol, polyvinyl alcohol, polyvinylpyrrolidone, hydroxypropyl methylcellulose, methylcellulose, carboxymethylcellulose and hydroxyethyl cellulose.
4. Surfactant : Benzthonium chloride, polysorbate-20 etc.
5. Hydrogen ion concentration : The greatest comfort for the patient is expected to be found at the normal pH approximately 7.4, of lachrymal fluid but usually adequate in the pH range of 6 to 8. Most of the active ingredients used in ophthalmic solution are salts of weak acids and have low buffer capacity e.g. sodium acid phosphate solution, sodium phosphate solution, sodium chloride solution etc.
6. Buffers : Monobasic sodium phosphate solution, dibasic sodium phosphate solution and sodium chloride solution.
7. Tonicity modifiers : sodium chloride and boric acid.
8. Antioxidant : Solution metabisulphite(0.1%) and sodium thiosulphite (0.1%)
9. Chelating agent : Disodium edetate
10. Antimicrobial : Benzylkonium chloride, chlorahedine acetate, chlorbutol, thromersal, phenylethyl alcohol, PMN and PMA.

11. Clarity : All solutions for the use in the eye should be clear and free from foreign particles.
12. Sterilization : Ophthalmic solution are required to be sterile when prepared and great care must be exercised subsequently to prevent contamination to use. The microbe must frequently found as a contaminant is *Pseudomonas aeruginosa*. The solution most often found contaminant is of sodium fluerescein. It is very dangerous and opportunistic organism that rows well on most culture media and produces both toxin and anti-bacterial products. This Gram -ve neative bacillus also rows readily in ophthalmic solutions which may becomes the source of extremely serious infections of the cornea. The common methods of sterilization are moist heat under pressure (autoclave), bacterial filtration (pore size 0.22 μm) and use of chemicals.

CONTAINERS

1. Single dose containers : It is minim unit and consists of a disposable, pliable, tube-like applicator made from polypropylene by ultrasonic sealing. It has a nozzle protected by a cap. It is generally enclosed in a heat-sealed envelope.
2. Multiple application containers : The traditional container for eye drops is a glass bottle carrying in the cap a dropper fitted with a teat. e.g.
   a. Amber colored container to provide protection from light
   b. Hexagonal in cross-section and vertically ribbed on three adjacent faces, the ribs indicated that the contents must not be taken orally.
   c. Made from neutral or treated soda glass.
   d. Screw capped bottle with separate droppers.

LABELLING

Comply with the general requirements for labelling, following information is mentioned on the label clearly.

1. Name and concentration of medicaments
2. Name and concentration of antimicrobial
3. Date after which the product is not intended to be used
4. Discard the contents at a specified time after opening (about one month)
5. If any discomfort or irritation discontinue use the eye drop
6. 'For External use Only;

HOW TO USE EYE DROPS

1. Wash hands with soap
2. Pull lower eyelid down with one hand gently

3. If dropper is separate, squeeze rubber bulb once while dropper is in bottle to bring liquid into dropper.
4. Holding dropper above eye, drop medicine inside lower lid while lookin up.
5. Do not touch dropper to eye or fingers
6. Lift lower lid and try to keep eye open
7. Do not blink for at least 30 seconds
8. If dropper is separate, replace on bottle and tighten the cap
9. Dropper tip always down
10. Never rinse the dropper
11. Never use drops that have changed color
12. Open one bottle at a time if you have more than one bottle at a time
13. If you are using more than one kind of drop at the same time, wait several minutes before use other drop
14. After instillation of drops, do not close eyes tightly and try not to blink more often than usual.

EYE LOTIONS

Eye lotions are sterile aqueous solutions containing no bactericide for first aid purposes over a maximum periods of twenty four hours. In another words, eye lotions are aqueous solutions containing a bactericide for intermittent domiciliary administration for up to seven days. Eye lotions are usually applied with an eye bath, which for first aid purposes only hence it has limited use nowadays but may be prescribed. e.g. sodium bicarbonate eye lotion is used for the emergency treatment of acid burns of the eye and is popular as 'Factory eye drops No.2', sodium chloride is an isotonic solution which is occasionally used for irrigating the eye. Soothing lotions for tired eyes.

CONTAINERS

Eye lotions are dispensed in colored fluted bottle closed with a screw-cap. It is important to ensure that the liner of the screw cap is not covered by cork as it may be a source for microbial contamination. Rubber or plasties liners are suitable and the screw cap may be either of metal or plastics. Uncolored for non-fluted bottles may be used if a suitable colored, fluted bottle is not available.

LABELLING

Following information is state on the eye lction label

1. 'For External use Only'
2. Discard any lotion remaining after use when the seal is first broken

If the eye lotions which are issued for domiciliary purposes should be labelled to indicate

1. Not to be used more than 24 h. (without bactericide)
2. Not to be used more than 7 days (with bactericide)

EYE OINTMENTS

Ophthalmic ointments are ointments meant for application to the eye. It can be used to obtain the effect of a variety of medicaments on the outside and edges of the eyelids, the conjunctiva, the cornea and the iris. It contains sterilized ingredients packed under rigidly aseptic conditions and meets the requirements of the official sterility tests. Ophthalmic ointments must contain a suitable substance or mixture of substance to prevent growth of or to destroy, microorganisms accidentally introduced when the container is opened during use. The medicinal agent is added to the ointment base either as a solution or as a micronized powder.

The finished ointment must be free from large particles. Most ophthalmic ointments are prepared with a base of white petroleum and mineral oil often with anhydrous lanolin.

CONTAINER

1. Ointments are packed in collapsible metal or plastic tubes or in single dose containers
2. Plastic tubes are used for some proprietary formulations of eye ointments
3. Certain proprietary formulations of eye ointments are available in single dose containers, elongated, flexible gelatin capsule with one end constructed and are opened by cutting off the constructed end with sterile scissors.

LABELLING

The label on the tube or outer sealed envelop should state that the contents are sterile provided that the container has not been opened. Other information should be similar as discussed for eye preparations.

EXERCISE NO. 1

Object : To prepare solution for eye drops

Requirements : Methyl hydroxybenzoate, propyl hydroxybenzoate, purified water, aseptic room, autoclave, and common glasswares.

Theory : Eye drops must be freshly prepared aseptically and dispensed in previously sterilized containers. A suitable fungistatic should be used in preparations liable to support the growth of moulds. For oily eye drops the oily vehicle, which has been previously sterilized by heating at 160°C for one hour must be used.
Eye drops should be made approximately isotonic with lachrymal secretion by the addition of sodium chloride or other suitable substance. Care should be taken to avoid contamination during use.

Formula

Methyl hydroxybenzoate	22.0 mg
Propyl hydroxybenzoate	11.4 mg
Purified water, sufficient to produce	100.0 ml

Procedure

1. Dissolve methyl hydroxybenzoate and propyl hydroxybenzoate in boiling water U under aseptic condition.
2. Add freshly boiled and cooled purified water to produce the required volume.

Category : Pharmaceutical aid

Storage : Preserve in a well-closed container and protect from light and contamination,

Uses : Pharmaceutical aid

VIVA VOCE QUESTIONS

1. How will you prepare eye drop?
2. What is difference between eye drops and eye lotion?
3. Give pharmaceutical application of eye drops.
4. Name of preservatives used in the eye drops preparation.

EXERCISE NO. 2

Object : To prepare Zinc sulphate eye drop

Requirements : Zinc sulphate, phenylmercuric acetate, purified water and other aseptic room facility and common glasswares.

Theory : Zinc sulphate eye drops are sterile solution of zinc sulphate containing phenyl mercuric acetate or phenylmercuric nitrate as an antimicrobial agent. Zinc sulphate is colorless, transparent crystals, or crystalline powder, astringent and metallic. It is efflorescent in dry air.

Formula

Zinc sulphate	12.5 mg
Sodium chloride	40.0 mg
Solution for eye drops, q.s. to	5.0 ml

Procedure

1. Dissolve weighed amount of zinc sulphate in purified water with aseptic precautions.
2. Dissolve phenylmercuric acetate in small amount of purified water
3. Add this solution and mix if required filter it aseptically.

Category : Astringent

Dose : One or two drops

Storage : Preserve in a well-closed container

Uses : Astringent

VIVA VOCE QUESTIONS

1. What is eye drops?
2. What is pharmaceutical application of zinc sulphate?
3. Give examples of anti-microbial agents.
4. What is astringent?
5. How eye drops are sterilized?

EXERCISE NO. 3

Object : To prepare Zinc sulphate and adrenaline eye drops

Requirements : Zinc sulphate, adrenaline acid tartrate, phenylmercuric acetate or phenylmercuric nitrate, sodium metabisulphite and purified water.

Formula

Zinc sulphate	250.0 mg
Adrenaline acid tartrate	90.0 mg
Phenylmercuric acetate or Phenylmercuric nitrate	2.0 mg
Sodium metabisulphite	50.0 mg
Purified water, sufficient to produce	100.0 ml

Procedure

1. Dissolve zinc sulphate, adrenaline acid tartrate in 50.0 ml of purified water with aseptic condition.
2. Dissolve phenylmercuric acetate and sodium metabisulphite in small amount of purified water.
3. Add this solution in sulphate solution and shake it. If necessary filter it.

Category : Astringent

Dose : use one or two drops

Storage : Preserve in a well-closed container

Uses : Astringent

VIVA VOCE QUESTIONS

1. What are eye drops?
2. How eye drops are sterilized?
3. Give principle of autoclave.
4. How to validate the instrument of sterilization?
5. What is fumigation?
6. What is aseptic transfer?

EXERCISE NO. 4

Object : To prepare Amethocane eye drops

Requirements : Amethocaine hydrochloride or tetracaine hydrochloride, sodium chloride, solution for eye drops, aseptic room and common glasswares.

Formula

Amethocaine hydrochloride	50.0 mg
Sodium chloride	34.0 mg
Solution of eye drops, qs to	5.0 ml

Procedure

1. Dissolve sodium chloride in solution of eye drop and prepare a solution.
2. Add amethocaine hydrochlorde and dissolve it by shaking.
3. Sterile by autoclave.

Category : Local anesthetic

Dose : 0.5 ml solution to the conjunctiva, repeated at 5 to 10 min. interval if necessary.

Storage : Preserve in a well-closed container (sterile preparation)

Uses : Local anesthetic, topically on the eye and by infiltration for spinal anesthesia.

VIVA VOCE QUESTIONS

1. What is amethocaine hydrochloride?
2. How will you prepare amethocaine hydrochloride?
3. What is topical anaesthetic?
4. How this preparation can be sterilized?
5. What is the use of eye drop?

EXERCISE NO. 5

Object : To prepare Atropine sulphate eye drops

Requirements : Atropine sulphate, sodium chloride, solution for eye drops, aseptic room and common glasswares

Formula

Atropine sulphate	50.0 mg
Sodium chloride	37.5 mg
Solution for eye drops, q.s. to	5.0 ml

Procedure

1. Dissolve sodium chloride in solution of eye drop and prepare a solution.
2. Add atropine sulphate and dissolve it by shaking.
3. Sterile by autoclave.

Category : Antimuscarinic

Dose : 1 to 2 drops

Storage : Preserve in a well-closed container.

Uses : For acute inflammation of the anterior tract. The half-life in the eye is long, and effects may lasts for 7 to 12 days after topical application to the eye.

VIVA VOCE QUESTIONS

1. What is the use of atropine sulphate?
2. What is difference in eye drops and ear drops?
3. Give storage condition of eye drops.

EXERCISE NO. 6

Object : To prepare Cocaine hydrochloride eye drops

Requirements : Cocaine hydrochloride, sodium chloride, solution for eye drops etc.

Formula

Cocaine hydrochloride	100.0 mg
Sodium chloride	30.0 mg
Solution for eye drops, q.s. to	5.0 ml

Procedure

1. Dissolve sodium chloride in solution of eye drop and prepare a solution.
2. Add cocaine hydrochloride and dissolve it by shaking.
3. Sterile by autoclave.

Category : Local anaesthetic **Dose :** 2 to 3 drops

Storage : Preserve in a suitable container. **Uses :** Local anaesthetic

VIVA VOCE QUESTIONS

1. What is cocaine?
2. Why eye drops are required?
3. What is local anaesthetic

EXERCISE NO. 7

Object : To prepare Fluorescein eye drops

Requirements : Fluorescein sodium, sodium chloride, purified water , aseptic room etc.

Formula

Fluorescein sodium	100.0 mg
Sodium chloride	16.0 mg
Purified water, q.s. to	5.0 ml

Procedure

1. Dissolve sodium chloride in purified water and shake. If necessary filter it.
2. Add fluorescein sodium and dissolve it.
3. Sterilize by autoclaving.

Category : Diagnostic aid. **Dose :** Topical, 0.1 to 0.3 ml of 2% solution

Storage : Preserve in a well closed container.

Uses : Diagnostic aid and to the conjunctiva.

VIVA VOCE QUESTIONS

1. What is the use of eye drops of fluorescein sodium?
2. How eye drops are prepared?
3. Give sterilization parameters.
4. What is conjunctiva?

EXERCISE NO. 8

Object : To prepare Framycetin eye drops

Requirements : Framycetin sulphate, sodium chloride, solution for eye drops, etc.

Formula

Framycetin sulphate	25.0 mg
Sodium chloride	37.5 mg
Solution for eye drops, q.s. to	5.0 ml

Procedure

1. Dissolve sodium chloride in solution of eye drop and prepare a solution.
2. Add Framycetin sulphate and dissolve it by shaking.
3. Sterile by autoclave.

Category : Anti-microbial **Dose :** 2 to 3 drops

Storage : Preserve in a well-closed container.

Uses : Anti-microbial

VIVA VOCE QUESTIONS

1. What is eye drops?
2. Which types of dosage form is more suitable for eye?
3. What is eye disk?

4. What is the capacity of cornea?

EXERCISE NO. 9

Object : To prepare Homatropine eye drops

Requirements : Homatropine hydrochloride, sodium chloride, solution for eye drops.

Formula

Homatropine hydrochloride	100.0 mg
Sodium chloride	28.0 mg
Solution for eye drops, q.s. to	5.0 ml

Procedure

1. Dissolve sodium chloride in solution of eye drop and prepare a solution.
2. Add homatropine hydrochloride and dissolve it by shaking.
3. Sterile by autoclave.

Category : acute inflammation **Dose :** 2 to 3 drops

Storage : Preserve in well-closed container. **Uses :** acute inflammation

VIVA VOCE QUESTIONS

1. Which types of diseases are generally found in the eye?
2. Eye drop is a sterile preparation or not.
3. Why eye drops should be sterile?
4. Which types of microorganism is more susceptible to the eye?

EXERCISE NO. 10

Object : To prepare Physostigmine eye drops

Requirements : Physostigmine salicylate, sod. chloride, sodium solution for eye drops.
Formula

Physostigmine salicylate	25.0 mg
Sodium chloride	40.0 mg
Sodium metabisulphate	2.0 mg
Solution for eye drops, q.s. to	5.0 ml

Procedure

1. Dissolve sodium chloride in small amount of solution for eye drops.
2. Add physostigmine salicylate and shake.
3. Dissolve sodium metabisulphate in solution for eye drops.
4. Sterile the product.

Category : Anticholinesterases **Dose :** 2 to 3 drops in affected eye.

Storage : Preserve in a well closed container. **Uses :** Anticholinesterases

VIVA VOCE QUESTIONS

1. What is the use of alkaloid?
2. What is the pharmacological action of alkaloids
3. Give storage conditions.

EXERCISE NO. 11

Object : To prepare Pilocarpine eye drops

Requirements : Pilocarpine nitrate, sodium chloride, solution for eye drops, etc.

Formula

Pilocarpine nitrate	50.0 mg
Sodium chloride	34.0 mg
Solution for eye drops	5.0 ml

Procedure

1. Dissolve sodium chloride in solution of eye drop and prepare a solution.
2. Add pilocarpine nitrate and dissolve it by shaking.
3. Sterile by autoclave.

Category : glaucoma **Dose :** 2 to 3 drops

Storage : Preserve in a well closed container protected from light.

Uses : glaucoma

VIVA VOCE QUESTIONS

1. What is glaucoma?
2. How to manage glaucoma?
3. What is the use of pilocarpine?

EXERCISE NO. 12

Object : To prepare Silver nitrate eye drops

Requirements : Pilocarpine nitrate, sodium chloride, solution for eye drop, etc.

Formula

Silver nitrate	25.0 mg
Sodium chloride	65.0 mg
Solution for eye drop	5.0 ml

Procedure

1. Dissolve sodium chloride in solution of eye drop and prepare a solution.
2. Add Silver nitrate and dissolve it by shaking.
3. Sterile by autoclave.

Category : Antiseptic, disinfectant and astringent

Dose : 2 to 3 ml

Storage : Preserve in a well-closed container.

Uses : Antiseptic, disinfectant, caustic and astringent

VIVA VOCE QUESTIONS

1. What is antiseptic?
2. What is disinfectant?
3. What is difference between antibacterial and antiseptic?

EXERCISE NO. 13

Object : To prepare Sulphacetamide eye drops

Requirements : Sulphacetamide

Formula

Sulphacetamide sodium	1.5 g
Solution of eye drops, q.s. to	5.0 ml

Procedure

1. Dissolve sulphacetamide sodium in solution of eye drops.
2. Add sufficient solution of eye drops to produce 5.0 ml.
3. Sterilize the product.

Category : Antibacterial **Dose :** 2 to 3 drops

Storage : Preserve in a well-closed container. **Uses :** Antibacterial

VIVA VOCE QUESTIONS

1. What is sulphonamide?
2. What is the pharmacology of sulphonamide?
3. What is the mechanism action?

EXERCISE NO. 14

Object : To prepare Prednisolone eye drops

Requirements : Prednisolone, sodium chloride, phenyl alcohol, solution of eye drops.

Formula

Prednisolone	0.5 mg
Sodium chloride	37.5 mg
Sphenyl alcohol	25.0 mg
Solution of eye drops	5.0 ml

Procedure

1. Dissolve sodium chloride in solution of eye drops.
2. Mix sphenyl alcohol with prednisolone.
3. Mix both solution and sterilize.

Category : Glucocorticoids **Dose :** 2 to 3 drops

Storage : Preserve in a well-closed container. **Uses :** as glucocorticoids

VIVA VOCE QUESTIONS

1. What is the use of eye drops?
2. How will you prepare eye drops?

EXERCISE NO. 15

Object : To prepare Sodium chloride eye lotion

Requirements : Sodium chloride, purified water etc.

Formula

Sodium chloride	0.9 g.
Purified water, sufficient to produce	100.0 ml

Procedure

1. Dissolve the sodium chloride in the purified water.
2. Clarify by filtration
3. Transfer the solution in a well-closed container
4. Sterile by autoclave

Category : Pharmaceutical aid **Storage :** Preserve in a well-closed container.

Uses : Pharmaceutical aid

VIVA VOCE QUESTIONS

1. What is eye lotion?
2. What is the use of eye lotion?
3. How to prepare eye lotion?

EXERCISE NO. 16

Object : To prepare Simple eye ointment

Requirements : Wool fat, yellow soft paraffin, liquid paraffin, etc.

Theory : Eye ointments must be free from large particles and should be prepared with aseptic precautions. For the preparation of simple eye ointment, it is often necessary to vary the proportion of the different ingredients of the bases to maintain a suitable consistency under different climatic conditions. Liquid paraffin, white soft paraffin and hard paraffin may be adjusted for this purpose. However, the proportions of the active medicaments must not be altered.

Formula

Wool fat	10.0 g
Yellow soft paraffin	80.0 g
Liquid paraffin, sufficient to produce	100.0 g

Procedure

1. Melt together weighed amount of wool fat and yellow soft paraffin in a container.
2. Add liquid paraffin to make up weight 100 g.
3. Filter the hot mixture through coarse filter paper placed in a heated funnel.
4. Filtrate is sterilized by dry heat at 150°C for sufficient time to ensure that the whole is maintained at this temperature for one hour.
5. Allow to cool at room temperature without opening the container.

Category : Pharmaceutical aid.

Storage : 5.0 g in small sterilized collapsible tube with the stated strength should be dispensed unless otherwise directed. Eye ointment may be unstable and should be stored in a cool place.

Uses : Pharmaceutical aid.

VIVA VOCE QUESTIONS

1. How to prepare base for eye ointment?
2. What is the composition of simple eye ointment?
3. Why ointment are prepared?
4. How one can change the composition of eye ointment.
5. How will you sterilize ointment?

EXERCISE NO. 17

Object : To prepare Chloramphenicol eye ointment

Requirements : Chloramphenicol, wool fat, yellow soft paraffin, liquid paraffin, etc.

Formula :

Chloramphenicol	0.2 g
Simple eye ointment	10.0 g

Procedure :

1. Triturate chloramphenicol with simple eye ointment in aseptic conditions
2. Add sufficient quantity of sterile simple eye ointment.

Category : Antibiotic

Storage : Preserve in a well-closed container in cool place.

Uses : Anti-microbial or infection

VIVA VOCE QUESTIONS

1. How to prepare chloramphenicol eye ointment?
2. What is the use of ointment?
3. Why ointment should be stored in cool place?

Note : Other eye ointments are prepared by the same method as chloramphenicol, containing different concentration of drug in simple eye ointment such as
Framycetin eye ointment (0.5%), Hydrocortisone eye ointment (2.5%), Physostigmine eye ointment (0.125%), Pilocarpine eye ointment (1.0%), Sulphacetamide eye ointment (6.0%), and Tetracycline eye ointment (1.0%).

13

Parenterals

- Water for injection
- Ascorbic acid injection
- Dextrose injection
- Sodium chloride injection
- Calcium gluconate injection
- Chloramphenicol injection
- Cyanocobalamin injection
- Dextrose with sodium chloride solution
- Electrolyte Maintenance IV fluid
- Electrolyte Maintenance IV fluid solution
- Gastrointestinal electrolyte maintenance IV fluid
- Sodium chloride hypertonic injection
- Compound sodium chloride injection
- Sodium citrate anticoagulant injection
- Compound sodium lactate injection

Injections are sterile products intended for administration of drugs by injection under or through one or more layers of skin or mucous membrane. The parenteral routes of drug administration are indicated for one or more of following reasons

1. Less amount of drugs required because adequate concentrations of the drug reaches at the site of action.
2. It controls some pharmacological parameters such as time of drug onset, serum peak level, tissue concentration and rate of elimination of drug from the body.
3. Useful when other routes are not available for administration of drug.
4. It also provides a local effect when it is desirable to minimize or avoid systemic toxic effects or reactions.
5. It is helpful in administered of drugs to unconscious, uncooperative or uncontrolled patients.
6. It exhibits biological effect that can not be achieved through oral administration because of nonabsorbance of from the alimentary canal.
7. It controls correction of fluid and electrolyte imbalance rapidly.
8. It minimizes side effect of drug in the body.

TERMINOLOGY

1. **Ampul :** An all-glass hermetically sealed container for a parenteral or other sterile product.

2. **Aseptic :** It means 'without sepsis', a condition in which every reasonable means has been used to destroy or eliminate viable microorganisms but this condition is not absolutely achieved.

3. **Clean :** It is relative term designing freedom from contamination. When used with parenteral medications 'clean' normally designates highly efficient cleanliness level.

4. **Large volume parenteral** (LVP) : A liquid intended for infusion and hermetically sealed in a container of greater than 100 ml volume.

5. **Small volume parenteral** (SVP) : A pareteral preparation hermetically sealed in a container of 100 ml or less volume.

6. **Multiple-dose vial :** A container usually of glass, having a relatively large opening closed by means of a rubber stopper which permits the insertion of a sharp needle and withdrawing a part of its contents.

7. **Nosocomial :** It is infection associated with a hospital.

8. **Pyrogen :** It is lipid associated with a polysaccharide or polypeptide of microbial origin and produces fever.

Types of Injections

1. Solutions ready for injection
2. Dry, soluble products ready to be combined with a solvent for injection.
3. Suspensions ready for injection
4. Dry insoluble products ready to be combined with a vehicle just prior to use.
5. Emulsion

Quality Control

Following quality control tests are carried out for parenteral dosage form.

1. Leaker test
2. Clarity test
3. Pyrogen test
4. Sterility test

ROUTES OF ADMINISTRATION

Route	Description
I. Primary route	
1. Intramuscular	Directly into the body of a relaxed muscle
2. Intravenous	Directly into a vein
3. Subcutaneous	Into the loose connective and adipose tissue beneath the skin
II. Secondary route	
1. Intra-arterial	Into an artery which leads directly to the target organ.
2. Intraabdominal	Directly into the peritoneal cavity via needle or inwelling catheter or directly into an abdominal organ such as the kidney or bladder
3. Intra-articular	Into synovial sacs of various accessible joints
4. Intracardiac	Directly into chambers of the heart.
5. Intracisternal	Directly into the cisternal space surrounding the bas of the brain
6. Interadermal	Into the dermis located just beneath and adjacent to the epidermis
7. Intralesional	Directly into or around a lesion, usually located in or on the skin or soft tissues to achieve a therapeutic effect.
8. Intraocular	Four types of intraocular injections are utilized
a. Anterior chamber	Directly into the anterior chamber of the eye.
b. Interavitreal	Directly into the vitreous cavity of the eye.
c. Retrobulbar	Around the posterior segment of the globe.
d. Subconjunctival	Injection are given beneath the conjunctiva.
9. Intrapleural	Usually single injection into the pleural cavity.
10. Intrathecal	Directly into lumbar sac located at the caudal end of spinal cord.
11. Intrauterine	Injection via a needle inserted percutaneously into pregnant uterus.
12. Intraventricular	Directly into the lateral ventricular of the brain.
13. Hypodermolysis	Subcutaneous route of administration for infusion of large volume

FORMULATION

Additives	Examples
1. Vehicles	
a. Aqueous	Water for injection, Sterile water for injection, etc.
b. Non-aqueous	PEG 400,600, propylene glycol, ethyl alcohol, dioxolanes, dimethyacetamide, ethyl oleate, isopropyl myristate, etc.
2. Antimicobial agents	Benzyl alcohol, benzthonium chloride, butylparaben, methylparaben, propylparaben, phenylmercuric nitrate, thimerosal, chlorobutanol, phenol, etc.
3. Solubilizing agents	PEG-300, polysorbate 20,40 and 80, povitone, ethyl alcohol, ethyl lactate, glycerin, eggyolk, dimethylacetamide, etc
4. Buffers	Sodium acetate, sod. bicarbonate, sod. carbonate, sod. citrate, sod. tartrate, actic acid, adipica acid, etc
5. Tonicity modifiers	Lactose, mannitol, sorbitol, dextrose, sodium chloride, etc.
6. Suspending agents	PEG 4000, pectin, gelatin, sorbitol solution, sodium CMC,
7. Antioxidants	Ascorbic acid, citric acid, tartaric acid, phosphoric acid, thiourea tocopherol, sod. bisulphite, sod. metabisulfite, BHT,
8. Chelating agents	Edetate disodium, edetate calcium disodium, etc
9. Stabilizers	Niacinamide, glycine, creatinine, sod.caprylate, etc.
10. Local anesthetics	Procaine hydrochloride, benzyl alcohol

Small Volume Parenterals (SVP) : Small volume of parenterals include ampouples of 1 ml, 2 ml, 3 ml, 5, ml upto 20 ml and vials of 1 ml, 2 ml, upto 30 ml. They are administered by various routes such as intramuscular, intravenous, subcutaneous, intraspinal, intracisternal, or intrathecal etc.

Large Volume Parenterals (LVP) : LVP contain 100 ml and more of injectable solution. They are generally administered by intravenous route. Two types of solutions are available in the market.

a. Electrolytes : sodium chloride and potassium chloride solutions, etc
b. Non-electrolytes : Dextrose and mannitol solutions, etc

Powder Parenterals : Dry powders are available in the market.. They are soluble in water for injection or in another sterile solvent just prior to use. They are available in form of vials containing rubber closure and sealed with aluminium file. Following antibiotics are in the form of dry powder. e.g. penicillin, ampicillin, chloramphenicol, amoxycillin, sterptomycin, etc.

STERILIZATION

Injections are sterilized by following methods :

1. Moist heat sterilization
2. Dry heat sterilization
3. Filtration method using bacteria proof filters
4. Using aseptic techniques

Container Materials

Material	Water vapor permeation	Gas permeation	Reaction with product	Physical properties
I. Thermoplastic Polymers				
1. Polyethylene				
a. Low density	High	Low	Low	Translucent, flexible
b. High density	Low	Low	Low	Translucent, semi-rigid
2. Polypropylene	Moderate	Low	Low	Translucent, semi-rigid
3. Polyvinyl chloride				
a. Flexible	High	Low	Moderate	Transparent, flexible
b. Rigid	High	Low	Low	Transparent, rigid
4. Polycarbonate	High	Low	Low	Transparent, rigid
5. Polyamide	High	Low	High	Translucent, rigid, tough
6. Polystyrene	High	High	Moderate	Transparent, rigid
7. Teflon	Low	Low	Nil	Translucent, rigid, tough temperature resistant
II Glass				
1. Soda-lime	None	None	High	Optically clear, rigid
2. Borosilicate	None	None	Low	Optically clear, rigid
II. Rubber				
1. Butyl	Low	Moderate	Moderate	Opaque, flexible
2. Natural	Moderate	Moderate	High	Opaque, flexible
3. Neoprene	Moderate	Moderate	High	Opaque, flexible
4. Polyisoprene	Moderate	Moderate	Moderate	Opaque, flexible
5. Silicone	Very high	Very high	Low	Translucent, flexible

Volume of Injection

Each single dose container contains slightly excess amount that can not be withdrawn. The excess volumes usually permitted are shown in the following table.

S.No.	Volume specified (ml)	Excess volume permissible (mobile liquid) (ml)	Excess volume permissible (viscous liquid) (ml)
1.	0.5	0.10	0.12
2.	1.0	0.10	0.15
3.	2.0	0.15	0.25
4.	5.0	0.30	0.50
5.	10.0	0.50	0.70
6.	20.0	0.60	0.90
7.	50.0 or more	about 2%	about 3%

Cleaning and Sterilization of Vials

Vials are cleaned and sterilized before filling. Following procedure is used for cleaning and sterilization of vials.

1. Soak the vial with detergent solution for overnight to make it free from sticking particles, grease, etc.
2. Wash with tap water three to four times till soap solution is completely removed.
3. Remove surface alkalinity using 1.0% hydrochloric acid solution.
4. Again wash the vial using tape water till free from alkalinity.
5. Rinse with deionized water and finally with distilled water.
6. Cleaned vials are sterilized by dry heat sterilization at 200°C for 4 hours.
7. Cool sterilized vials at room temperature under the closed condition prior to filling the vials.

Cleaning and Sterilization of Rubber closures

1. Rubber closures are boiled with 1.0% solution of liquid detergent for 30 minutes.
2. Wash with tap water till free from detergent.
3. Boil for 30 min. using 1.0% solution of hydrochloric acid.
4. Wash to make them free from acid.
5. Boil the acid washed rubber closures with 1.0% sodium carbonate solution and wash till free alkali.
6. Wash till free from alkalinity.
7. Treat rubber stoppers with double strength of bacteriostatic solution.
8. Wash three to four times using pyrogen free water.
9. Sterilize by autoclave at 115°C for 30 minutes.

EXERCISE NO. 1

Object : To prepare Water for injection

Requirements : Portable water

Theory : Water for injection can not be prepared by simple distillation method because in that case the pyrogens may enter mechanically into the condenser and ultimately into the receiver. Injection of distilled water may cause rise in body temperature due to the presence of pyrogens. Water, which is increasing temperature of body, is known as apyrogenic.

It can also cause chills, pains in back and legs. Pyrogen are metabolite end products of bacteria. They are non-filterable, thermostable, and non-volatile in nature. The pH of water for injection should be as near as possible to pH 7.0.

It should also be free from carbon dioxide because slight acidity of water for injection is sufficient to precipitate the alkaloidal salts like phenobarbitone and sulphonamides. Water for injection should be free from dissolved air. It is necessary to prevent sensitive medicament from oxidation.

Formula

Portable water q.s.	100.0 ml

Procedure

1. Distilled portable water from a neutral glass or metal still fitted with an efficient device from preventing the entrapment of droplets.
2. Reject the first portion of distillate.
3. Collect the remaining amount of water in suitable container.
4. Immediately sterilize by heating in a autoclave or by filtration without addition of a bacteriostatic.

Category : Pharmaceutical aid

Storage : Preserve in a well-closed container and maintain aseptic conditions

Uses : For the preparation of water for injection and pharmaceutical aid.

EXERCISE NO. 2

Object : To prepare Ascorbic acid injection

Requirements : Ascorbic acid, sodium bicarbonate, propylene glycol, sodium hydrosulphite, disodium edetate, water for injection, etc.

Formula

Ascorbic acid,	1.03 g
Sodium bicarbonate	2.47 g
Propylene glycol	2.50 g
Sodium hydrosulphite	0.25 g
Disodium edetate	0.05 g
Water for injection, sufficient to produce	50.0 ml

Procedure

1. Dissolve ascorbic acid slowly under constant stirring and nitrogen bubbling.
2. Add sodium bicarbonate and slow stirring.
3. Separately prepare solution of sodium hydrosulphite in small proportion of water for injection and add in the solution.
4. Add solution of disodium edetate and mix uniformly.
5. Add mix propylene glycol with slow stirring.
6. pH is adjusted to 5.5 to 6.5 using sodium hydroxide solution or ascorbic acid solution.
7. Make up the volume and mix with water for injection.
8. Sterilized by passing through sterilized 0.22 µ membrane filter into sterilized filling vessel.
9. Aseptically fill 5.3 ml in sterilized 5 ml amber ampoules.

Category : Vitamin

Dose : 100 mg

Storage : Preserve in a well-closed container

Uses : Deficiency of vitamin

EXERCISE NO. 3

Object : To prepare Dextrose injection

Requirements : Dextrose, water for injection, etc.

Theory : Dextrose solution is a sterile solution of anhydrous dextrose or an equivalent quantity of dextrose monohydrated for parenteral solution. If the concentration of the solution is not reported, a 5% w/v solution should be prepared because it is isotonic with blood serum. The dextrose decomposes on heating. The rate of decomposition depends on sterilization temperature, duration, of sterilization and presence of other substances. Decomposition of the dextrose is reduced by adjusting the pH between 3.5 to 6.5 using hydrochloric acid and heating time should be avoided as possible as and over heating should be avoided. It should be cooled as quickly as possible after sterilization of the solution. Solution should not be adhere to the neck of ampoules because it is converted in black form during the process of sealing. Injection should not be used if it contain a precipitate or any suspended particle.

Formula

Dextrose	5.0 g
Water for injection	100.0 ml

Procedure

1. Dissolve the weighed quantity of dextrose in water for injection.
2. Add sufficient water for injection to produce 100 ml.
3. If necessary filter the solution.
4. 10 ml of solution is filled into each ampoule.
5. The ampoules are sealed and sterilized immediately by autoclave.

Category : Source of energy
Dose : As per condition of the patient. The maximum dose is 12 liters per meter body surface per 24 hours by IV drip.

Storage : Preserve in a hermetically sealed ampoules.

Uses : It is used as source of energy. It is used in the treatment and prevention of toxic effects of drugs on liver. In fever and other conditions when nitrogen excretion is increased its use prevents loss of nitrogen. Mixture with sodium chloride is used in severe nutritional disturbances of infants particularly when they suffer from diarrhoea and vomiting.

Note : Dextrose injection can also be sterilized by filtration under aseptic conditions.

EXERCISE NO. 4

Object : To prepare Sodium chloride injection

Requirements : Sodium chloride, water for injection, etc.

Formula

Sodium chloride	0.9 g
Water for injection, sufficient to produce	100.0 ml

Procedure

1. Dissolve sodium chloride in sufficient quantity of water for injection.
2. Add sufficient water for injection to produce 100 ml.
3. If necessary filter the solution.
4. Fill the solution into the ampoules and seal them.
5. Sterilize by autoclave as early as possible.
6. Sterilization can also be done by filtration.

Category : Electrolyte

Storage : Preserve in a hermetically sealed container.

Uses : It is used in electrolyte deficiency

EXERCISE NO. 5

Object : To prepare Calcium gluconate injection

Requirements : Calcium gluconate, calcium D-saccharate, water for injection, etc

Formula

Calcium gluconate	5.0 g
Calcium D-saccharate	1.75 g
Water for injection, sufficient to produce	100.0 ml

Procedure

1. Dissolve calcium gluconate in water for injection.
2. Add and dissolve calcium D-saccharate in solution.
3. Adjust the pH 6.0 to 8.2 using 10% solution of sodium hydroxide.
4. Filter through 0.45 μm membrane filter.
5. Fill in 10 ml ampoules and seal them.
6. Sterilize by autoclave at 121°C for 30 minutes.

Category : Fluid and electrolyte

Dose : Calcium gluconate by Intramuscular or intravenous injection 1 to 2 g.

Storage : Calcium gluconate injection is a supersaturated solution and must be completely free from solid particles. If solid particles are present and separation of crystals occurs, the injection must not be used. Preserve in a well-closed container.
Uses : Fluid and electrolyte replinisher

EXERCISE NO. 6

Object : To prepare Chloramphenicol injection

Requirements : Chloramphenicol, lignocaine hydrochloride, chlorocresol, distilled water, and propylene glycol, etc.

Formula

Chloramphenicol	12.5 g
Lignocaine hydrochloride	0.514 g
Lignocain base	0.405 g
Chlorocresol	1.0 g
Distilled water	12.0 ml
Propylene glycol	100.0 ml

Procedure

1. Add chloramphenicol in hot propylene glycol and stir till dissolved.
2. Add lignocaine base and dissolve using continuous stirring.

3. Dissolve chlorocresol and lignocaine hydrochloride separately in distilled water and add in the solution and stir well.
4. Add sufficient propylene glycol to produce sufficient volume upto 100 ml.
5. Adjust pH of the solution of 6.5 to 6.8.
6. Filter the solution by vacuum filtration using Whattman filter paper No.1 and then this solution is passed through G-3 sintered glass funnel.
7. Aseptically fill 10.5 ml of solution in dry sterile 10 ml amber vials with washed sterile butyl grey plugs and seal.

Category : Antibiotic

Dose : For an adult, 1.5 to 3.0 g daily in divided doses. For child, 25 to 50 mg per kg of body weight daily, in divided doses.

Storage : Preserve in a tightly-closed container, protected from light.

Uses : Antibiotic

EXERCISE NO. 7

Object : To prepare Cyanocobalamin injection

Requirements : Cyanocobalamin, sodium chloride, sodium dihydrogen phosphate, benzyl alcohol, water for injection, etc

Formula

Cyanocobalamin	0.12 g
Sodium chloride	0.75 g
Sodium dihydrogen phosphate	0.30 g
benzyl alcohol	1.0 ml
Water for injection, sufficient to produce	100.0 ml

Procedure

1. Dissolve cynocobalmin in water for injection.
2. Add benzyl alcohol and mix.
3. Add sodium chloride, and sodium dihydrogen phosphate separately in water for injection.
4. Mix the solution in the mixture of vitamins.

5. Add sufficient water to produce 100 ml.
6. Filter through G-3 sintered glass funnel
7. Adjust pH 4.0 to 5.5.
8. Fill aseptically in 10 ml amber vials.

Category : Vitamin

Storage : Preserve in a cool place.

Uses : Vitamin

EXERCISE NO. 8

Object : To prepare Dextrose with sodium chloride solution

Requirements : Dextrose, sodium chloride, activated charcoal, water for injection, etc

Formula

Dextrose	5.0%
Sodium chloride	0.9%
Activated charcoal	5.0%
Water for injection, sufficient to produce	100.0 ml

Procedure

1. Dissolve excess amount of dextrose in water for injection.
2. Add activated charcoal and other additives
3. Filter the solution.
4. Sterilize by autoclave.

Category : Nutrient

Storage : Preserve dextrose in cool place

Uses : Nutrient

Note : Bacteriostatic agents must not be added. If the ampoules show any color or contain a precipitate or any suspended impurities, it must be rejected.

EXERCISE NO. 9

Object : To prepare Electrolyte Maintenance IV fluid (for Paediatric use)

Requirements : Dextrose anhydrous, sodium acetate, potassium chloride, magnesium chloride, dibasic potassium phosphate, sodium metabisulphite, acetic acid, water for injection, etc.

Formula

Dextrose anhydrous	5.0%
Sodium acetate	0.315%
Potassium chloride	0.13%
Magnesium chloride	0.031%
Dibasic potassium phosphate	0.025%
Sodium metabisulphite	0.021%
Acetic acid	qs
Water for injection, sufficient to produce	100 ml

Procedure

1. Dissolve sodium acetate, potassium chloride, magnesium chloride, and dibasic potassium phosphate.
2. Adjust pH to 5.0 using glacial acetic acid.
3. Add dextrose anhydrous and sodium metabisulphite.
4. Make up the volume.
5. Again adjust pH between 4.8 to 5.0.
6. Filtration using 0.45 μm membrane filter.
7. Sterilize at 121°C for 30 min.

Category : Fluid and electrolyte replinisher

Storage : Preserve in a well-closed container and protected from light.

Uses : Fluid and electrolyte replinisher

EXERCISE NO. 10

Object : To prepare Electrolyte Maintenance IV fluid solution

Requirements : Dextrose anhydrous, sodium acetate, potassium chloride, sodium chloride, dibasic potassium phosphate, sodium metabisulphite, acetic acid, water for injection, etc.

Formula

Dextrose anhydrous	5.0%
Sodium acetate	0.28%
Potassium chloride	0.15%
Sodium chloride	0.09%
Dibasic potassium phosphate	0.13%
Sodium metabisulphite	0.02%
Acetic acid	qs
Water for injection, sufficient to produce	100 ml

Procedure

1. Dissolve sodium acetate, potassium chloride, sodium chloride, and dibasic potassium phosphate.
2. Adjust pH to 5.0 using glacial acetic acid.
3. Add dextrose anhydrous and sodium metabisulphite.
4. Make up the volume.
5. Again adjust pH between 4.8 to 5.0.
6. Filtration using 0.45 μm membrane filter.
7. Sterilize at 121°C for 30 min.

Category : Fluid and electrolyte replinisher

Storage : Preserve in a well-closed container, protected from light.

Uses : Fluid and electrolyte replinisher

EXERCISE NO. 11

Object : To prepare Gastrointestinal electrolyte maintenance IV fluid

Requirements : Dextrose anhydrous, sodium sulphite, potassium chloride, sodium chloride, ammonium chloride, hydrochloric acid, water for injection, etc.

Formula

Dextrose anhydrous	5.00 g
Sodium sulphite	0.015 g
Potassium chloride	0.13 g
Sodium chloride	0.37 g
Hydrochloric acid	q.s.
Ammonium chloride	0.37 g
Water for injection, sufficient to produce	100.0 ml

Procedure

1. Dissolve sodium chloride, potassium chloride, and ammonium chloride in water for injection stir till dissolved.
2. Adjust pH to 4.8 to 5.0 with dilute hydrochloric acid.
3. Add dextrose anhydrous and sodium sulphite.
4. Again adjust the pH between 4.8 to 5.0 with glacial acetic acid.
5. Filtration using 0.45 μm membrane filter.
6. Sterilize at 121°C for 30 min.
7. Fill 100 ml in 540 me glass bottles for IV fluids.

Category : Gastrointestinal electrolyte

Storage : Preserve in a well-closed container, store in cool place.

Uses : Gastrointestinal electrolyte

EXERCISE NO. 12

Object : To prepare Sodium chloride hypertonic injection

Requirements : Sodium chloride, water for injection, etc.

Formula

Sodium chloride	1.60 g
Water for injection, sufficient to produce	100.0 ml

Procedure

1. Dissolve sodium chloride in water for injection.
2. Add sufficient water for injection to produce 100 ml.
3. Filter the solution and immediately sterilize by autoclave.
4. It can be sterilized by filtration.

Category : Fluid and electrolyte replinisher

Dose : The dose is determined by the prescriber in accordance with the needs of the patient.

Storage : Sodium chloride hypertonic injection on keeping may cause the separation of small particles from glass containers. A solution containing such particles must not be used.

Uses : Fluid and electrolyte replinisher

EXERCISE NO. 13

Object : To prepare Compound sodium chloride injection

Synonyms : Ringer's Injection

Requirements : Sodium chloride, potassium chloride, calcium chloride hydrated, water for injection, etc.

Formula

Sodium chloride	0.86 g
Potassium chloride	0.030 g
Calcium chloride hydrated	0.033 g
Water for injection, sufficient to produce	100.0 ml

Procedure

1. Dissolve sodium chloride, potassium chloride, calcium chloride hydrated in water for injection.
2. Add sufficient water for injection to produce 100 ml.
3. Filter the solution and immediately sterilize by autoclave.
4. It can be sterilized by filtration.

Category : Fluid and electrolyte replenisher.

Dose : The dose is determined by the prescriber in accordance with the need of the patient.

Storage : Sodium chloride hypertonic injection on keeping may cause the separation of small particles from glass containers. A solution containing any particle must not be used.

Uses : Fluid and electrolyte replenisher.

EXERCISE NO. 14

Object : To prepare Sodium citrate anticoagulant injection

Requirements : Sodium citrate, water for injection, etc

Formula

Sodium citrate	4.0 g
Water for injection, sufficient to produce	100.0 ml

Procedure

1. Dissolve the sodium citrate in 90 ml of water for injection.
2. Filter and add sufficient water for injection to produce 100 ml.
3. Immediately sterilize by heating in an autoclave or by filtration.

Category : Anticoagulant

Storage : Sodium citrate anticoagulant injection may cause the separation of small solid particles from a glass container. A solution containing such particles must not be used. Sodium citrate anticoagulant injection kept in a container which is closed with cotton wool is used within one month after its preparation.

Uses : Anticoagulant for plasma and for blood for fractionations.

EXERCISE NO. 15

Object : To prepare Compound sodium lactate injection

Requirements : Lactic acid, sodium hydroxide, sodium chloride, calcium chloride, potassium chloride, dilute hydrochloric acid, water for injection, etc.

Formula

Lactic acid	0.24 ml
Sodium hydroxide	0.114 g
Sodium chloride	0.60 g
Calcium chloride	0.04 g
Potassium chloride	0.04 g
Dilute hydrochloric acid	q.s.
Water for injection, sufficient to produce	100.0 ml

Procedure

1. Dissolve the sodium hydroxide in 20 ml of water for injection.
2. Add lactic acid and dissolve it.
3. Heat in an autoclave at 115° to 116° for one hour.
4. Cool and add dilute hydrochloric acid until a few drops of solution give an orange color with solution of phenol red.
5. Dissolve other additive in separate flask.
6. Mix these two solutions.
7. Add sufficient purified water to produce 100 ml.
8. Filter and immediately sterilize by autoclave or by filtration.

Category : Fluid and electrolyte replenisher

Storage : Compound sodium lactate injection on keeping may cause the separation of small solid particles from a glass container. A solution containing such particles must not be used.

Uses : Fluid and electrolyte replenisher

VIVA VOCE QUESTIONS

For Exercise No. 1 to 15

1. What are parenterals?
2. What are injections?
3. What is infusion?
4. What is aseptic technique?
5. How glasswares, powder drug and thermosensitive solutions are sterilized?
6. How route of administration of drug is decided?
7. Give ideal requirements for the preparation of injections.
8. Define SVP and LVP.
9. What is membrane filter?
10. Why parenteral dosage form is required?
11. How to seal ampoules?
12. What are quality control parameters of the parenterals?

14

Powders

- Bulk oral powder
- Divided power
- Effervescent granules
- Aromatic chalk powder
- Compound bismuth oral powder
- Codeine compound powder
- Compound magnesium trisilicate oral powder
- Kaolin compound powder
- Salicylic acid compound dusting powder
- Zinc oxide and starch dusting powder
- Zinc undecylenate dusting powder
- Compound tragacanth powder
- Liquorice compound powder
- Mercury with chalk powder
- Barium sulphate powder
- Compound sodium chloride and dextrose oral powder

Historically, powders are one of the oldest dosage forms. A pharmaceutical powder is a mixture of finely divided drugs or chemicals in dry forms. They are intended for internal or external use for many diseases. Most of powders as a dosage form have been replaced largely by capsules and tablets.

Advantages

1. Rapid dispersion of ingredients
2. Good chemical activity
3. Easy to prepare
4. Economic

Disadvantages

1. Dose inaccuracy
2. Not suitable for dispensing of many unpleasant testing, hygroscopic and deliquescent substances.
3. Time consuming process.
4. Poor stability of powder

Classification of Powder

Powders are subdivided solids which are classified in the BP according to the size of their constituent particles of range from 1.25 μm to 1.7 mm in diameter. Another classification of powder is based on the dispensing of powder.

1. Bulk powder for external use e.g. dusting powders, douche powders, dentrifies, and insufflation.
2. Bulk powder for internal use.
3. Divided powder for internal use.
4. Effervescent granules
5. Cachets
6. Hard capsules.

The degree of coarseness or fineness of a powder is differentiated and expressed by reference to the nominal mesh aperture size of the sieves used. The term used in the classification of powders are

1. Coarse powder : All particle passing through sieve with a nominal mesh aperture of 1.70 mm and not more 40% through a sieve mesh aperture of 355 μm.
2. Moderately coarse powder : All particle passing through sieve with a nominal mesh aperture of 710 μm and not more 40% through a sieve mesh aperture of 250 μm.
3. Moderately fine powder : All particle passing through sieve with a nominal mesh aperture of 355 μm and not more 40% through a sieve mesh aperture of 180 μm.
4. Fine powder : All particle passing through sieve with a nominal mesh aperture of 180 μm.
5. Very fine powder : All particle passing through sieve with a nominal mesh aperture of 125 μm.

Sieve : Sieve should be made up of wires of uniform circular cross section. Wires may be of stainless steel or of other suitable material except that plated wires are not allowed. Wire mesh sieves are used in sifting powered drugs. They are identified by numbers indicating the nominal mesh aperture.

Nominal mesh aperature size (mm)	App. sieve number	Nominal mesh aperature size (μm)	App. sieve number	Nominal mesh aperature size ((μm)	App. sieve number
4.00	4	710	22	150	100
3.35	5	600	25	125	120

2.80	6	500	30	106	150
2.00	8	425	36	90	170
1.70	10	355	44	75	200
1.40	12	300	52	63	240
1.18	14	250	60	53	300
1.00	16	180	85	45	350

Preparation of Powders

On large scale preparation of powders, various mills are used such as rotary cutter, hammer, roller and fluid energy mill. On a small scale preparations of powder following communication techniques are employed.

1. Trituration : Particle size is reduced by rubbing it in a mortar with pestle.
2. Levigation : Particle size is reduced by adding a suitable nonsolvent known as levigating agent to form a paste and then mixed in mortar with pestle. This method is used to incorporate solids into ointments and suspension to remove gritty feel.
3. Pulverization by intervention : Particle size is reduced by adding a suitable material that can be removed easily after pulverization. This method is most suitable for gummy substances.

Following methods are employed for mixing of powders :

1. Spatulation : Small amount of powders are mixed by the movement of a spatula on a sheet of paper or a pill tile.
2. Trituration : suitable for comminute and to mix powders
3. Geometric dilution or mixing : It is suitable when potent substances are to be mixed a large amount of diluent. The potent dug is placed upon an approximately equal volume of the diluent in mortar and mixed by trituration. A second portion of diluent equal in volume to the powder mixture in mortar is added, and trituration is repeated. (e.g. Potent drug- 1 g, Diluent – 15 g, mix in such a way 1+1 = 2 + 2 = 4 + 4 = 8 + 8 = 16 , 1 is the quantity of potent drug and other quantity represents diluent)
4. Sifting : Powders are mixed by passing them through sifters. This process is suitable for light and fluffy product but not suitable for potent drugs.
5. Tumbling : It is a process of mixing the powder in a large container rotated by a electric motor.

Bulk powders are dispensed in bulk container. Divided powders are packed in powder papers. Powder papers are of four types.

1. Vegetable parchment, a thin semi-opaque moisture resistant paper.
2. White bond, an opaque paper with no moisture-resistant properties.
3. Glassine, a glazed, transparent moisture-resistant paper.
4. Waxed, a transparent waterproof paper.

Special Problems

1. Volatile substances : The loss menthol, camphor, and essential oils are lost by volatilization when mixed in powders. The volatility may be reduced or retard by use of heat- sealed plastic bags or by double wrapping with a waxed or glassine papers inside the bond papers.

2. Eutectic mixtures : Liquids result from the combination of some substances during milling or mixing of powders are known as eutectic mixtures such as phenol, camphor, salol, menthol, phenacetin, antipyrine, thymol, aspirin, acetanilid, etc. Inert substances are used to retard this problem.

3. Hygroscopic and deliquescent substances : Some substances become moist because of affinity for moisture in the air. Powders may be prepared by adding inert diluents. Deliquescent substances can not be satisfactorily prepared as powders.

4. Liquids : In divided powder liquids may be incorporated in small amounts. Magnesium carbonate, starch or lactose may be added to increase the absorbability of the powder if necessary.

EXERCISE NO. 1

Object : To prepare Bulk oral powder

Synonym : Gregory's powder

Requirements : Heavy magnesium carbonate, light magnesium carbonate, rhubarb in powder, ginger in powder, etc.

Theory : Gregory's powder is rhubarb powder. It contains the light and heavy varieties of magnesium carbonate. In general, the light form is better for liquid medicines because it disperses easily and remains longer in suspension while he heavy form, because it is less bulky. It is more satisfactory for powders

Formula

Heavy magnesium carbonate	325 g
Light magnesium carbonate	325 g
Rhubarb in powder	250 g
Ginger in powder	100 g

Procedure

1. Incorporate the powders in order of bulk, adding at each addition, a quantity that approximately doubles the bulk already in the mortar.
2. Mix properly and uniformly.
3. Take a mortar that comfortably holds all of the mixture and do not grind too vigorously because heavily rubbed powders become compacted and less easily miscible with water.
4. Loosen the powder from the bottom of the mortar and scrape it from the sides. To scrape quickly use a large flexible spatula.
5. Pass the powder through sieve 250 and then remix lightly with a large spatula.

Category : Bulk oral powder

Storage : Powder must be preserve in a light-resistant container to prevent discoloration of the rhubarb and container should be store in a airight because ginger and rhubarb have volatile constituents.

Uses : Bulk oral powder

VIVA VOCE QUESTIONS

1. What are powders?
2. What is the synonym of rhubarb powder?
3. What is bulk powder?
4. What is the use of bulk powder?
5. What is the use of ginger in the powder?
6. Give advantages and disadvantages of powder.
7. How to prepare powder?

EXERCISE NO. 2

Object : To prepare Divided powder

Synonym : Seidliz powders

Requirements : Sodium potassium tartrate, sodium bicarbonate and tartaric acid, etc

Theory : In 1815, T.F.Savory was introduced the Seidlitz powder. It represents the principle constituent of the Seidlitz spring in Czechoslovakia. In BPC, it is still under the name of effervescent powder compound. It consists of two powders.

1. Powder A contains sodium potassium tartrate and sodium bicarbonate and is wrapped in blue paper.
2. Powder B contains tartaric acid and is wrapped in white paper.

It is used as an effervescent draught. The contents of blue paper (Powder A) are dissolved in a tumblerful of cold or warm water, the contents of the white paper (Powder B) are added, the mixture is stirred and the product is taken while effervescing. Preparations are more agreeable when slightly acid because the salty taste is less noticeable. This powder contains 12% more tartaric acid than is necessary to neutralize the sodium bicarbonate. When an increase in strength is desired, the amount of sodium potassium tartrate is raised to twice as much, in effervescent powder, compound double strength.

Formula

Powder A

Sodium potassium tartrate	7.5 g
Sodium bicarbonate	2.5 g

Powder B

Tartaric acid	2.5 g

Procedure

1. Mix sodium potassium tartrate and sodium bicarbonate uniformly.
2. Similarly powder tartaric acid separately.
3. Double wrap both powders, using an inner wrapper of wax paper.
4. Wrap both powder in different color papers.
5. Pack in pairs should consists of powder A and powder B.

Category : Effervescent draught

Dose : Take one pair of powder at a time.

Storage : Both sodium bicarbonate and tartaric acid are hygroscopic in nature at high humidity. Hence both powder should be stored at low humidity.

Uses : Effervescent draught

VIVA VOCE QUESTIONS

1. What is the use of effervescent powder?
2. Why both powder are packed in separate packet?
3. What will happen if powder B is, added first in water?
4. How to take this powder?
5. What is draught?
6. What is the use of tartaric acid?
7. How to enhance the activity of powder?
8. Give pharmaceutical application of powder.

EXERCISE NO. 3

Object : To prepare Effervescent granules

Requirements : Sodium bicarbonate, citric acid, tartaric acid, sucrose, etc.

Theory : Effervescent granules contain following substances:
Sodium bicarbonate : This reacts with the acid when the preparation is added to water. The carbon dioxide produces the effervescence. The effervescence from the release of carbon dioxide serves to mask the taste of salty or bitter medications.

Citric acid and tartaric acids : The quantity of these acids is slightly higher than is necessary to neutralize the sodium bicarbonate because effervescent preparations are more palatable if slightly acid. The relative proportions of the two acids is based on experience of the amount of water needed to make the mixture sufficiently coherent but citric acid has one molecule of water of crystallization equal to 8.75% of weight. When it is heated it liberates one molecule of water and the moist condition thus produced allows partial interaction between the acids and the bicarbonate.

The water of crystallization of the citric acid and the water from the reactions makes the material coherent. Loss of weight occurs during granulation due to the following reasons.

(a) Evaporation from the damp mixture.
(b) Loss of carbon dioxide.

It was observed that the losses constitute approximately one-seventh of the weight of powder used and must be allowed for when calculating the amount to be prepared.

Chemical reaction

$$3\ NaHCO_3 + C_6H_8O_7.H_2O \ = \ C_6H_5Na_2O_7 + 3\ CO_2 + 3\ H_2O$$

$$2\ NaHCO_3 + C_4H_6O_6 \ = \ C_4H_4Na_2O_6 + 2\ CO_2 + 2\ H_2O$$

Formula

Sodium bicarbonate, in powder	4.08 g
Tartaric acid, in powder	2.16 g
Citric acid, in powder	1.55 g
Refined sugar, in powder	1.20 g
Iron and ammonium citrate, in powder	0.43 g

Procedure

1. Place a porcelain dish on a water-bath with the dish deep in the bath and heat the bath to boiling point.
2. Powdered weighed amount of substance and mix.
3. Place all the powders in the hot porcelain dish on a boiling water-bath.
4. Stir the mixture by a bone spatula or scoop until the mixture has formed either a loose cake or a damp coherent mass.
5. Mix the mixture continuously.
6. Sieve no. 8 superposed upon a no. 20 or 24 and quickly press the mixture through the top sieve and then shake. The finest granules will drop through the lower sieve, leaving upon it practically uniform granules which constitute the required product.
7. Spread the granules out on a tray in a warm place and dry for 2 or 3 hours.
8. Dry granules are kept in a suitable container.

Category : Effervescent granules

Dose : As prescribed

Storage : Dry granules are sent in bulk in a wide-mouth bottle, or separately wrapped in double powder papers and labeled properly.

Uses : Antacid

VIVA VOCE QUESTIONS

1. What are effervescent granules?
2. Why granules are prepared?
3. How store granules?
4. What is the use of two acids?
5. Why citric acid is used in effervescent granules?
6. What is antacid?
7. What is the use of effervescent granules?
8. Define effervescent powder, granules and tablets.
9. Give storage condition of the tablets.

EXERCISE NO. 4

Object : To prepare Aromatic chalk powder

Requirements : Chalk, cinnamon, nutmeg , clove, cardamom seed, sucrose, etc.

Formula

Chalk	0.015 g
Cinnamon, finely powdered	60.0 mg
Nutmag, finely powdered	48.0 mg
Clove, finely powdered	24.0 mg
Cardamom seed freshly removed from capsule and finely powdered	18.0 mg
Sucrose	0.30 g

Procedure

1. Mix all substances in ascending order of its weight and triturate uniformly.
2. Pass through suitable sieve and preserve in a well-closed container.

Storage : It should be kept and dispensed in air-tight containers.

Uses : Carminative and antacid.

VIVA VOCE QUESTIONS

1. What is compound powder?
2. What is the use of chalk?
3. What is carminative?
4. What is the use of various additives in the powder?

EXERCISE NO. 5

Object : To prepare Codeine compound powder

Synonym : APC powder

Requirements : Aspirin, Phenacetin, Codeine phosphate, and general glasswares.

Formula

Aspirin	0.25 g
Phenacetin	0.25 g
Codeine phosphate	0.08 g

Procedure

1. Separately powder a slight excess of the additives which are crystalline form.
2. Weigh out the required amount of each powder and of lactose, if necessary. Dilute the ingredient of smallest amount with the others, using them in ascending order of weight and mixed uniformly.
3. Weigh out the prescribed number of powders and wrap it properly.

Category : Analgesic

Storage : Preserve in a well-closed container.

Uses : Analgesic

VIVA VOCE QUESTIONS

1. What is APC?
2. How to prepare APC powder?
3. What is the use of Phenacetin?
4. What is the use of codeine phosphate?
5. Give storage condition of powder.

Note : Always weigh excess quantity of ingredients than required (at least equivalent to one powder). Weigh the surplus material, which should roughly equal the theoretical excess. If the difference is large it may indicate inaccurate weighing, and the exercise should be repeated.

EXERCISE NO. 6

Object : To prepare Compound Magnesium trisilicate oral powder

Synonym : Compound Magnesium trisilicate powders

Requirements : Magnesium trisilicate, Chalk, heavy magnesium carbonate, sodium bicarbonate, etc.

Formula

Magnesium trisilicate	0.25 g
Chalk in powder	0.25 g
Heavy magnesium carbonate	0.25 g
Sodium bicarbonate	0.25 g

Procedure

1. Mix additives in mortar and pestle and shake it uniformly.
2. Dispensed in a suitable container.

Category : Antacid **Dose :** 0.3 to 2.0 g.

Storage : Preserve in a well closed container. **Uses :** Antacid

VIVA VOCE QUESTIONS

1. Why magnesium trisilicate powder is prepared?
2. What is antacid?
3. What is the use of magnesium carbonate?

EXERCISE NO. 7

Object : To prepare Kaolin compound powder

Requirements : Light kaolin, heavy magnesium carbonate, sodium bicarbonate, etc.

Formula

Light kaolin	1.00 g
Heavy magnesium carbonate	0.67 g
Sodium bicarbonate	0.33 g

Procedure : Same as exercise no. 6.

Category : Diarrhoea **Storage :** Preserve in a well-closed in container.

Uses : Diarrhoea

VIVA VOCE QUESTIONS

1. What is light kaolin?
2. What is diarrhoea?
3. How to prepare pharmaceutical powders?
4. Give advantages of powder?
5. What is the use of magnesium carbonate?

EXERCISE NO. 8

Object : To prepare Talc dusting powder

Requirements : Magnesium oxide, starch, purified sterilized talc, etc

Formula

Magnesium oxide	0.50 g
Starch in powder	1.00 g
Purified sterilized talc	10.0 g

Procedure : Same as exercise no. 6.

Category : Pharmaceutical aid, dusting powder

Storage : Preserve in a well-closed container in a wide-mouthed bottle.

Uses : Pharmaceutical aid, dusting powder

VIVA VOCE QUESTIONS

1. What are dusting powder?
2. What is the use of purified talc?
3. How dusting powder is prepared?
4. Give application of dusting powder.

EXERCISE NO. 9

Object : To prepare Salicylic acid compound dusting powder

Requirements : Salicylic acid, boric acid, purified talc, etc

Formula

Salicylic acid, in powder	0.30 g
Boric acid , in powder	0.50 g
Purified talc, sterilized	10.0 g

Procedure : Same as exercise no. 6

Category : Germicide, local anti-infective

Storage : Preserve in a well-closed container. **Uses :** Germicide, local anti-infective

VIVA VOCE QUESTIONS

1. What is the use of boric acid?
2. What is germicide?
3. What is local anti-infective?
4. What is dusting powder?
5. How dusting powder is prepared?
6. Classify powders with examples.
7. Which type of containers is suitable to preserve dusting powder?

EXERCISE NO. 10

Object : To prepare Zinc oxide and starch dusting powder

Requirements : Zinc oxide, starch, etc.

Formula

Zinc oxide	5.00 g
Starch	5.00 g

Procedure : Same as exercise no. 6.

Category : Astringent, protective and antiseptic

Storage : Preserve in a well-closed container.

Uses : Astringent, protective and antiseptic

EXERCISE NO. 11

Object : To prepare Zinc undecylenate dusting powder

Requirements : Zinc undecylenate, Undecenoic acid, turpentine oil, starch, light kaolin, etc.

Formula

Zinc undecylenate	5.0 g
Undecenoic acid	1.0 g
Turpentine oil	0.25 ml
Starch	25.0 g
Light kaolin	50.0 g

Procedure : Same as exercise no. 6

Category : Antifungal agent

Storage : Preserve in a well-closed container.

Uses : Antifungal agent, it is only fungistatic not fungicidal.

VIVA VOCE QUESTIONS

Exercise No. 11 & 12

1. What are antifungal?
2. What is fungistatic?
3. What is difference between fungistatic and fungicidal?
4. How to prepare powder?

EXERCISE NO. 12

Object : To prepare Compound Tragacanth powder

Requirements : Tragacanth, acacia, starch, sucrose, etc.

Theory : Tragacanth is a gum which is obtained from various species of *Astragalus gummifer*. It is used as a suspending agent, granulating agent and thickening agent. Acacia is a dried gum obtained from stem and branches of *Acacia Arabia*. It is used in many pharmaceutical preparations as a binding agent, emulsifying agent, granulating, suspending agent, etc. Starch is obtained from potato, wheat, rice, pea, and maize. The application of the starch depends on the form of use. It is commonly used as suspending agent, granulating agent, disintegrating, binding and thickening agent. Sucrose is obtained from sugar cane. It is used as sweetening agent, binding agent and its USP syrup used as a preservative because 85% solution of sucrose has high osmotic pressure.

Formula

Tragacanth, in powder	1.50 g
Acacia, in powder	2.00 g
Starch, in powder	2.00 g
Sucrose, in powder	4.50 g

Procedure

1. Mix weighed amount of substances in ascending order of its weight.
2. Mix properly and store in a suitable container.

Category : Pharmaceutical aid

Storage : The powder is stored in a well-closed container and protected from atmospheric moisture.

Uses : Pharmaceutical aid as suspending agent, thickening agent, emulsifying agent etc.

VIVA VOCE QUESTIONS

1. What is demulcent?
2. What is the use of tragacanth?
3. How will you prepare powder?
4. What is use of acacia?

EXERCISE NO. 13

Object : To prepare Liquorice compound powder

Requirements : Senna leaf, Liquorice, fennel, sulphur, sucrose, etc

Theory : Senna consists of dried leaflets of *Cassia senna* and it contains not more than 1.0 percent of its stalk. It contains anthraquinone derivatives which are laxative in nature. Senna is also used as a purgative and original glycosides of reduced derivatives are more active than free anthrequinones. When the anthracene derivatives containing drugs are used alone they cause griping and thus they are usually administered with carminatives. Liquorice consists of dried peeled or unpeeled roots and stems of *Glycyrrhiza glabra*. It consists glycosides and has demulcent or soothing action. Fennel consists of the dried ripe fruits of *Foeniculum vulgare*. It is used as a respiratory stimulant and is used as an expectorant. It is an aromatic, carminative and flavoring agent and is generally added in cough and digestive mixtures. Due to sweet taste anehole is used in mouth and dental preparations. In this preparation sulphur is converted into alkali sulphide, which have irritant action and produces a mild laxative action. Finely powder means a powder, which can pass through sieve no. 85. Sieve number 85 is defined as the sieve which has 85 meshes or wires of standard thickness in one inch length or 2.54 cm. The main medicament in this powder is senna leaf. It is used in constipation. This product is prescribed for persons suffering from habitual constipation.

Formula

Senna leaf, in powder	1.60 g
Liquorice, peeled, in powder	1.60 g
Fennel, in powder	0.80 g
Sulphur, sublimed	0.80 g
Sucrose, in powder	5.20 g

Procedure

1. Mix weighed amount of substances in ascending order of the weight used in the formulation of the powder.
2. Mix properly and store in a suitable container.

Category : Laxative **Dose :** 4 to 8 g

Storage : Preserve in a well closed container and protected from atmospheric moisture and to prevent the evaporation of volatile oil.

Uses : Laxative (Laxative are mild purgative)

VIVA VOCE QUESTIONS

1. What is liquorice?
2. What is senna?
3. What is fennel?
4. How this powder is prepared?
5. Define carminative with examples.
6. Define laxative, purgative and cathartic.

EXERCISE NO. 14

Object : To prepare Mercury with chalk powder

Synonym : Grey powder

Requirements : Mercury, dextrose, chalk, etc.

Theory : Mercury is finely divided with chalk powder which is chemically calcium carbonate. In trituration dextrose helps in distribution of mercury. This preparation is rarely used as mercury is poisonous in nature if taken more than the prescribed quantity of powder. It is essential to preserve the preparation in an air-tight container to avoid oxidation of mercury. Oxidised product of mercury is more poisonous than mercury. Higher temperature for storage is not suitable because some times mercury can coalesce together and separated in the form of globules.

Formula

Mercury	3.30 g
Dextrose	0.3 g
Chalk	6.60 g

Procedure

1 Weighed amount of mercury, dextrose and chalk mix in ascending order in a mortar and triturates with pestle.
2 Mix the mixture until acquires a uniform pale grey color and has no metallic globules when examine by the microscope.

Category : Purgative **Dose :** 60 to 300 mg

Storage : Preserve in a well-closed container.

Uses : Purgative

VIVA VOCE QUESTIONS

1. Why powder of mercury is prepared?
2. What is purgative?
3. Why this powder is becoming obsolete?
4. What is the use of chalk?

EXERCISE NO. 15

Object : To prepare Barium sulphate powder

Synonym : Shadow meal or Barium meal

Requirements : Barium sulphate, saccharin sodium, vanillin, etc.

Formula

Barium sulphate	100.0 g
Saccharin sodium	0.025 g
Vanillin	0.010 g

Procedure

1. Mix weighed amount of substances in ascending order of its weight.
2. Mix properly and store in a suitable container.

Category : Diagnostic agent (Radio-opaque medium)

Dose : Orally 120 to 300 g in suitable suspension

Storage : Preserve in a well-closed container.

Uses : It is used as a contrast medium for X-ray examination of alimentary tract. The powder is mixed immediately before use with a sufficient quantity of water.

VIVA VOCE QUESTIONS

1. What is barium powder?
2. Classify dusting powder.
3. Define bulk powder.

EXERCISE NO. 16

Object : To prepare Compound sodium chloride and dextrose oral powder

Synonym : Electrolyte powder

Requirements : Sodium chloride, potassium chloride, sodium bicarbonate, dextrose, etc.

Formula

Sodium chloride, in powder	50.0 g
Potassium chloride, in powder	75.0 g
Sodium bicarbonate, in powder	75.0 g
Dextrose, in powder	10.0 g

Procedure

1. Mix weighed amount of substances in ascending order of its weight.
2. Wrap individual powder using an inner waxed paper or an outer covering of metal foil or dispersed in individual well-closed containers intended for use on one occasion.

Dose : In accordance with the needs and response of the patient.

Storage : Wrap individual powder using an inner wax paper or an outer covering of metal foil or dispersed in individual well-closed containers. The label on the container or the label on the package must state.

1. Direction for use
2. Any portion of the solution unused twenty four hours after preparation should be discarded.
3. Amount contained in the unit-dose container.

Uses : Used for electrolyte balance and dehydration.

VIVA VOCE QUESTIONS

1. What is electrolyte?
2. Which element is most essential for body?
3. Why electrolytes are prescribed by the doctors?
4. How to prepare electrolytes?

EXERCISE NO. 17

Object : To prepare Compound bismuth oral powder

Synonym : Compound bismuth powder

Requirements : Bismuth subcarbonate, sodium bicarbonate, calcium carbonate and heavy magnesium carbonate, etc.

Formula

Bismuth subcarbonate	1.25 g
Sodium bicarbonate	1.25 g
Calcium carbonate	3.75 g
Heavy magnesium carbonate	3.75 g

Procedure

1. Mix weighed amount of substances in ascending order of its weight.
2. Wrap individual powder using an inner waxed paper or an outer covering of metal foil or dispersed in individual well-closed containers intended for use on one occasion.

Category : Mild antacid, astringent, adsorbent, and protective

Dose : 1 to 5 g.

Storage : Preserve in a well-closed container.

Uses : Mild antacid, astringent, adsorbent, and protective

VIVA VOCE QUESTIONS

1. What is the use of bismuth subcarbonate?
2. What is chalk?
3. What is antacid?

15

Solutions

- Cresol with soap solution
- Aqueous Iodine solution
- Strong Iodine solution
- Weak Iodine solution
- Strong lead subacetate solution
- Dilute lead subacetate solution
- Strong ammonia solution
- Coal tar solution
- Chloroxylenol solution
- Strong coal tar solution
- Compound tartrazine solution
- Tolu solution
- Compound sodium chlcride solution

A solution is an homogeneous one-phase system consisting of two or more components. It contains two-phase i.e. solvent and solute. The solvent is the phase in which the dispersions occur and solute is that component which is dispersed as small ions or molecules in the solvents. In general solvent part is greater than solute in the solution except a few preparation. e.g. Syrup BP, it contains 66.7% of sucrose as solute and 33.3% of water as the solvents.

Advantages

1. Easy to swallow than solid dosage form like tablet and capsules.
2. Suitable for more paediatric preparation.
3. Drug in solution form is immediately available for absorption.
4. A solution is a homogeneous system and therefore the drug remains uniformly distributed throughout the preparation.
5. Drugs that can irritate and damage the gastric mucosa if localized in specific area, Its irritation is reduced by administration in solution form.

Disadvantages

1. In convenient to transport and store because they are bulky.
2. Whole product is lost immediately if any breakage in the container.
3. The stability of most of substances in aqueous solution is lesser than solid dosage form.
4. Shelf life of solution is shorter than solid preparation.
5. Suitable media for the microbial contamination and may therefore require the suitable preservatives.
6. Dose inaccuracy compared to solid dosage form.
7. Bitter unpleasant substances are not suitable for solutions and needs to add sweetening and flavoring agent to make them more palatable.

Formulation of Solution

Following additives are generally required for the preparation of solution.

1. Solvents : a. Aqueous b. Non-aqueous (fixed oil, alcohol, polyhydric alcohol, dimethyl-sulphoxide, ethyl ether, liquid paraffin, etc.)
2. Buffers : Carbonates, citrates, gluconates, lactates, phosphate, tartrate, borates, etc.
3. Colors : water soluble dye amaranth.
4. Density modifiers
5. Flavors and perfumes

Taste of product	Flavor
Salty	Apricot, butterscotch, liquorice, peach, vanilla
Bitter	Anise, chocolate, mint, wild cherry
Sweet	Vanilla fruits, etc.
Sour	Citrus fruits, raspberry, etc

6. Isotonicity modifiers
7. Preservatives
8. Antioxidants and reducing agents

CONTAINER

Narrow mouth, screw capped colorless plain bottle.

EXERCISE NO. 1

Object : To prepare Cresol with soap solution

Synonyms : Lysol

Requirements : Cresol, vegetable oil, potassium hydroxide, purified water, etc

Theory : Cresol is a mixture of o, m and p-cresol. It acts as disinfectant. The solubility of cresol in water is only about 3% whereas the quantity of cresol present in this preparation is 50%. For dissolving this high proportion of cresol, solubilising agent is required . Fatty acids containing vegetable oil react with hydroxide and form soap. This soap acts as solubilising agent. The reaction between vegetable oil and potassium hydroxide is known as saponification. The cresol is mixed only after completion of the reaction between vegetable oil and potassium hydroxide. The reaction between vegetable oil and potassium hydroxide can be enhanced by adding small amount of ethyl alcohol. Cresol is a phenolic compound and caustic in nature, therefore, it should be handled carefully.

Formula

Cresol	50.0 ml
Vegetable oil	18.0 g
Potassium hydroxide	4.20 g
Purified water, sufficient to produce	100.0 ml

Procedure

1. Dissolve potassium hydroxide in 25 ml of purified water.
2. Add vegetable oil and heat on a water-bath and mix thoroughly.
3. Continue heating until a small portion dissolves in water without separation of oily drops.
4. Add cresol and mix thoroughly.
5. Add sufficient purified water to produce the required volume.

Category : Disinfectant

Storage : Preserve in a narrow mouthed bottle

Uses : Disinfectant

VIVA VOCE QUESTIONS

1. What are solutions?
2. How cresol with soap solution is prepared?
3. What is disinfectant?
4. What is cresol?
5. What is the use of soap in this solution?
6. Give pharmaceutical applications of cresol with soap solution.
7. What is Lysol?

EXERCISE NO. 2

Object : To prepare Aqueous iodine solution

Synonyms : Lugol's solution

Requirements : Iodine, potassium iodide, purified water, etc.

Formula

Iodine	5.0 g
Potassium iodide	10.0 g
Purified water	100.0 ml

Procedure

1. Dissolve potassium iodide and iodine in 10 of purified water.
2. Add sufficient purified water to produce 100 ml.

Category : Source of Iodine **Dose :** 0.3 to 1.0 ml

Storage : Preserve aqueous Iodine solution in a well-closed container. It should be resistant to iodine.

Uses : Source of iodine

VIVA VOCE QUESTIONS

1. What is the use of iodine?
2. What is the use of potassium iodide in this solution?
3. Why iodine is used?

EXERCISE NO. 3

Object : To prepare Strong Iodine solution

Requirements : Iodine, Potassium iodide, purified water, ethanol, etc.

Formula

Iodine	10.0 g
Potassium iodide	6.0 g
Purified water	10.0 ml
Ethanol, sufficient to produce	100.0 ml

Procedure

1. Dissolve potassium iodide and iodine in purified water.
2. Add sufficient ethanol to produce 100 ml.

Category : Antiseptic

Storage : Preserve strong Iodine solution in a well-closed container. The material of the container should be resistant to iodine.

Uses : Antiseptic

VIVA VOCE QUESTIONS

1. How strong iodine solution is prepared?
2. What is difference between weak and strong iodine solution?
3. What is difference between antiseptic and disinfectant?
4. What is the use of ethanol in strong iodine solutions?
5. How to preserve weak iodine solution?

EXERCISE NO. 4

Object : To prepare Weak Iodine solution

Synonyms : Iodine Tincture

Requirements : Iodine, Potassium iodide, purified water, ethanol, etc

Formula

Iodine	2.0 g
Potassium iodide	2.5 g
Purified water	10.0 ml
Ethanol, sufficient to produce	100.0 ml

Procedure

1. Dissolve potassium iodide and iodine in purified water.
2. Add sufficient ethanol to produce 100 ml.

Category : Antiseptic **Dose :** 0.3 to 2.0 ml

Storage : Preserve strong Iodine solution in a well-closed container. The material of the container should be resistant to iodine.

Uses : Antiseptic

EXERCISE NO. 5

Object : To prepare Strong lead subacetate solution

Requirements : Lead acetate. Lead monoxide, purified water, etc.

Theory : Lead mono oxide and lead acetate react together and form a number of basic lead acetates. The basic lead acetate is formed by the following reaction.

$$(CH_3COO)_2Pb + PbO \rightarrow (CH_3COO)_2Pb.Pb(OH)_2$$

This basic acetate of lead is soluble in water. Highly alkaline solution of acetate of lead $(CH_3COO)_2Pb.Pb(OH)_2$ is poorly soluble.

Filtration should be conducted as quickly as possible and preferably performed in to a bottle in such a manner that the filtrate should not expose to carbon dioxide any longer than required because carbon dioxide of atmosphere precipitates the basic lead acetate and convert insoluble basic lead carbonate. For the same reason, the preparation is stored in small bottles which well filled and well closed.

Formula

Lead acetate	25.0 g
Lead monoxide, in powder	17.5 g
Purified water, sufficient to produce	100.0 ml

Procedure

1. Dissolve lead acetate in 75 ml of purified water.
2. Add lead monoxide.
3. Set a side for forty-eight hours.
4. Shaking, occasionally filter and pass through the filter.
5. Add sufficient purified water to produce 100 ml.

Category : Soothing astringent

Storage : Preserve in a well-filled and well closed container.

Uses : Soothing astringent

VIVA VOCE QUESTIONS

1. What is chemical reaction of lead acetate?
2. How will you prepare lead acetate solution?
3. Why this solution is prepared?
4. How will you differentiate solution and mixtures?
5. Give pharmaceutical application of lead acetate solutions.

EXERCISE NO. 6

Object : To prepare Dilute lead subacetate solution

Requirements : Lead acetate. Lead monoxide, purified water, etc.

Formula

Strong lead subacetate solution	1.25 ml
Purified water, sufficient to produce	100.0 ml

Procedure

It can be prepared in the manner as shown in Exercise No.5

Category : Soothing astringent

Storage : Preserve in a well-filled and well closed container.

Uses : Soothing astringent

EXERCISE NO. 7

Object : To prepare Strong ammonia acetate solution

Requirements : Glacial acetic acid, ammonium bicarbonate, ammonia solution strong, purified water, etc.

Theory : Strong solution of ammonium contains 57.5% w/v of ammonium acetate and it is prepared by reacting glacial acetic acid with ammonium bicarbonate and strong solution of ammonia.

$$NH_4HCO_3 + CH_3COOH \rightarrow (CH_3COO)\,NH_4 + H_2O + CO_2$$

$$NH_4OH + CH_3COOH \rightarrow (CH_3COO)\,NH_4 + H_2O$$

It is not possible to prepare the solution by reacting glacial acetic acid with ammonium bicarbonate alone because the reaction between these substances ceases before the formation of required amount of ammonium acetate. In this reaction the formation of ammonium acetate is enhanced and concentration of acetate ions increases and ionization of acetic acid decreases.

$$CH_3COOH \rightleftharpoons CH_3COO^- + H^+$$

$$NH_4HCO_3 \rightleftharpoons HCO_3^- + NH_4^+$$

$$H_4^+ + CH_3COO^- \rightleftharpoons CH_3COO\,NH_4$$

$$CH_3COO\,NH_4 \rightleftharpoons NH_4^+ + CH_3COO^-$$

Glacial acetic acid can not be replaced with acetic acid because glacial acetic acid is 99% pure while acetic acid is only 33% pure. This preparation can be prepared using ammonia strong solution alone because 453 g of glacial acetic acid is neutralized with about 1000 ml. Ammonium hydroxide is used to neutralize the unreacted acid and to make the preparation alkaline with pH range 7.6 to 8.1. The pH is adjusted by using two indicators like bromothymol blue and thymol blue.

Formula

Glacial acetic acid	45.3 g
Ammonium bicarbonate	47.0 g
Ammonia solution strong	10.0 g
Purified water, sufficient to produce	100.0 ml

Procedure

1. Mix the glacial acetic acid with about 35 ml of purified water.
2. Add the ammonium bicarbonate in small amount at a time until it is all dissolved.
3. Add sufficient of the ammonia solution until one drop of the resulting solution, diluted with ten drops of water.
4. Add one drop of solution of bromothymolblue, gives a full blue color and shows a full yellow color with one drop of solution of thymol blue.
5. Add sufficient purified water to produce the required volume.

Category : Diaphoretic **Dose :** 1 to 4 ml

Storage : Preserve in a well closed lead-free glass containers. **Uses :** Diaphoretic

VIVA VOCE QUESTIONS

1. What is the use of strong ammonia solution?
2. What do you mean by diaphoretic mixture?
3. How to prepare solution?
4. What are the chemical reactions involved in this preparation?
5. What is glacial acetic acid?

EXERCISE NO. 8

Object : To prepare Coal tar solution

Requirements : Prepared coal tar, quıllaia, ethanol, etc.

Formula

Prepared coal tar	20.0 g
Quillaia, in moderately coarse powder	10.0 g
Ethanol (90%), sufficient to produce	100.0 g

Procedure

1. Macerate the prepared coal tar and the quillaia with 80 ml of ethanol (90%) for seven days in a closed vessel, with occasional agitation.
2. Filter and add sufficient ethanol. Pass through the filter to produce 100 ml.

Category : Disinfectant

Storage : Preserve coal tar solution in a tight container.

Uses : Disinfectant

Note : In making coal tar solution the alcohol may be replaced by specially denatured spirit diluted so as to be equivalent alcoholic strength.

EXERCISE NO. 9

Object : To prepare Chloroxylenol solution

Requirements : Chloroxylenol, potassium hydroxide, oleic acid, castor oil, terpineol, ethanol (96%), purified water, etc.

Formula

Chloroxylenol	5.00 g
Potassium hydroxide	1.36 g
Oleic acid	0.75 ml
Castor oil	6.30 g
Terpineol	10.0 ml
Ethanol (96%)	20.0 ml
Purified water, sufficient to produce	100.0 ml

Procedure

1. Dissolve potassium hydroxide in small amount of purified water.
2. Add a solution of the castor oil in 63 ml of ethanol.
3. Mix and allow to stand for one hour or until a small portion of the mixture remains clear when diluted with nineteen times its volume of purified water and then add oleic acid.
4. Mix the terpineol with a solution of the chloroxylenol in the remainder of the ethanol and pour into the soap solution.
5. Add sufficient purified water to produce the required volume 100 ml.

Category : Topical antiseptic

Storage : Preserve in a well-closed container.

Uses : It is used as topical antiseptic in the treatment of minor burns, acne, eczema, psoriasis, seborrhea, diaper rash and in the treatment of athlete's foot.

Note : In making chloroxylenol solution, the ethanol may be replaced by industrial methylated spirit, provided that law and statutory regulations governing the use of Industrial methylated spirit are observed.

VIVA VOCE QUESTIONS

1. What is chloroxyenol?
2. What is use of potassium hydroxide in preparation of solution?
3. What is the use of terpineol in this preparation?
4. Give application of this solution.
5. Give storage and labelling condition of chloroxyenol solution.

EXERCISE NO. 10

Object : To prepare Strong coal tar solution

Requirements : Coal tar, polysorvate 80, ethanol (90%), etc.

Formula

Coal tar	40.0 g
Polysorvate 80	5.0 g
Ethanol (90%), sufficient to produce	100.0 ml

Procedure

1. Mix the coal tar and warm if necessary to render it fluid, with the polysorbate 80.
2. Pour this mixture in a thin stream into 70 ml of ethanol (96%) in a closed vessel fitted with an agitator.
3. Allow the mixture to stand for not less than twenty four hours.
4. Decant and filter the supernatant liquid, wash the vessel and filter with ethanol.
5. Combine the filtrate and washing and mix it.
6. Add sufficient ethanol to produce 100 ml.

Category : Pharmaceutical aid

Storage : Strong coal tar solution should be kept in a well-closed container.

Uses : Pharmaceutical aid

VIVA VOCE QUESTIONS

1. What is difference between weak and strong coal tar solutions?
2. What is the use of polysorbate 80 in the preparation of solutions?
3. Give pharmaceutical application of coal tar solution
4. What is the use of ethanol in the preparation of solutions?

5. How will you preserve coal tar solution?

EXERCISE NO. 11

Object : To prepare Compound tartrazine solution

Requirements : Tartrazine, sunset yellow FCF, glycerol, chloroform water, purified water, etc.

Formula

Tartrazine, food grade of commerce	0.75 g
Sunset yellow FCF, food grade of commerce	0.10 g
Glycerol	25.0 ml
Chloroform spirit	20.0 ml
Purified water, sufficient to produce	100.0 ml

Procedure

1. Dissolve the tartrazine and sunset yellow FCF in about 50 ml of the purified water.
2. Add glycerol, and mix uniformly.
3. Add chloroform spirit and purified sufficient to produce 100 ml.

Category : Pharmaceutical aid

Weight per ml : 1.040 to 1.060 g

Storage : Preserve in a well-closed container and protect from light

Uses : Pharmaceutical aid

VIVA VOCE QUESTIONS

1. What is tartrazine?
2. Why tartrazine solution is prepared?
3. How will you prepare solution?

EXERCISE NO. 12

Object : To prepare Tolu solution

Requirements : Tolu balsam, sucrose, ethanol, sterilized talc, purified water, etc.

Theory : Tolu Balsam is a solid or semi-solid, balsam obtained by incision from the trunk of *Myroxylon balsamum*. It contains not less than 35.0 percent and not more than 50.0 percent of total balsamic acids, calculated with reference to the dry alcoholic-soluble matter. It is soft, tenacious, brownish-yellow or brown mass, when first thin films. It has aromatic odor and vanilla like taste. It is used as expectorant.

Formula

Tolu balsam	5.0 g
Sucrose	50.0 g
Ethanol (90%)	30.0 ml
Purified water, sufficient to produce	100.0 ml

Procedure

1. Dissolve weighed quantity of tolu balsam in 200 ml of the ethanol.
2. Add about to 5% of previously sterilized purified talc and 35 ml of purified water.
3. Heat about to 70°C, shake vigorously.
4. Allow this mixture to stand for twenty four hours.
5. Filter the mixture and remove talc.
6. Dissolve sucrose in the filtrate.
7. Add remaining amount of ethanol and sufficient purified water to produce 100 ml.

Category : Expectorant **Weight per ml :** 1.400 to 1.1750 g

Dose : 2 to 8 ml **Storage :** Tolu solution should be kept in a well-closed container.

Uses : Expectorant

VIVA VOCE QUESTIONS

1. What is Tolu Balsam?
2. What is use of Tolu balsam?
3. How will you prepare tolu balsam solution?
4. What is expectorant?
5. What is use of ethanol in the preparation of solution?

EXERCISE NO. 13

Object : To prepare Compound sodium chloride solution

Synonyms : Ringer's solution

Requirements : Sodium chloride, potassium chloride, calcium chloride hydrated, purified water, etc.

Theory : It is an isotonic solution of Sodium chloride, potassium chloride, and calcium chloride. It is used as fluid and electrolyte replenisher to irrigate the tissues during experiments on animals. Ringer's injection has the same composition except water for injection in place of purified water, when it is used during cardiopulmonary bypass heart surgery.

Formula

Sodium chloride	8.6 g
Potassium chloride	0.3 g
Calcium chloride hydrated	0.33 g
Purified water, sufficient to produce	100.0 ml

Procedure

1. Dissolve sodium chloride, potassium chloride, and calcium chloride hydrated in a sufficient quantity of recently boiled purified water.
2. Add purified water to make up the volume.

Category : Fluid and electrolyte replenisher

Storage : Preserve in a well closed-container. Always freshly prepared solution is used.

Uses : Fluid and electrolyte replenisher

VIVA VOCE QUESTIONS

1. What is Ringer's solution?
2. How ringer solution is prepared?
3. Why used three salts are used in this solution?
4. What is the use of potassium chloride in the solutions?

16

Spirits

- Benzaldehyde spirit
- Lemon spirit
- Compound orange spirit
- Soft soap spirit
- Surgical spirit
- Chloroform spirits
- Spirits of ether
- Peppermint spirit
- Aromatic spirit of ammonia

Spirits are alcoholic or hydroalcoholic solution of volatile substances and contain 50% to 90% of alcohol. The high alcoholic contents maintain water-insoluble oils in solution. If water is added to a spirit the oil acts separated. Some spirits are used as flavoring agents e.g. compound orange spirit, cardamum spirits etc. and some spirits are used as medicinal agents e.g. aromatic ammonia spirit. Spirits are prepared by dissolving the volatile substances in ethanol 90 percent except aromatic ammonia spirit, prepared by distillation method.

CONTAINERS

Spirits should be kept in well-closed containers of glass or other suitable materials and should be stored at temperature not exceeding 25°C to reduce loss of evaporation.

LABELLING

Comply with the general requirements for labeling.

EXERCISE NO. 1

Object : To prepare Benzaldehyde spirit

Requirements : Benzaldehyde, ethanol, purified water and common glasswares

Theory : Benzaldehyde is a colorless, strongly refractive liquid and having an odor resembling that of bitter almond oil and a burning aromatic taste. It is also known as artificial essential almond oil. It is used in place of bitter almond oil for flavoring purposes. It is much safer than almond oil as it contains no hydrocynic acid. It is also used in perfumery and in manufacturing of dyestuffs and many other organic compounds.

Formula

Benzaldehyde	1.0 ml
Ethanol (90%)	80.0 ml
Purified water, sufficient to produce	100.0 ml

Procedure

1. Dissolve measured amount of benzaldehyde in the ethanol (90%).
2. Add sufficient purified water to make up the volume and mix properly by shaking.

Category : Flavoring agent

Weight per ml : 0.870 to 0.885 g

Storage : Benzaldehyde spirit should be kept in a well-filled container and protected from light.

Uses : Flavoring agent

VIVA VOCE QUESTIONS

1. What is spirit?
2. How spirits are prepared?
3. Why benzaldehyde spirit is needed?
4. Why it is necessary to store the preparation in a closed container?
5. Why are you prefer benzaldehyde than almond oil?
6. Give examples of other flavoring agent.
7. Define proof spirit and its importance in the pharmacy.

EXERCISE NO. 2

Object : To prepare Lemon spirit

Requirements : Terpeneless lemon oil, ethanol etc.

Theory : It is obtained from the fresh peel of fruit of *Citrus limon* Linn. by expression without heat with or without the previous separation of the pulp and the peel. Pale yellow to deep yellow or greenish yellow liquid with the characteristic odor and taste. Do not use lemon oil that has terebenthine odor because the terpenes of lemon oil is readily undergoes deterioration by oxidation, acquiring a terebinthinate odor. It is used a flavoring agent in pharmaceutical preparations and in certain candies and foods.

Formula

Terpeneless lemon oil	10.0 ml
Ethanol (90%), sufficient to produce	100.0 ml

Procedure

1. Mix measured amount of terpeneless lemon oil with ethanol (90%).
2. Add remaining volume of ethanol and make up the volume.

Category : Flavoring agent

Weight per ml : 0.814 to 0.823 g.

Storage : Lemon spirit should be kept in a well-filled container and protected from light.

Uses : Flavoring agent

VIVA VOCE QUESTIONS

1. What is lemon spirit?
2. How lemon oil gets deteriorated?
3. How to identify that oil is not suitable to use?
4. How to prepare lemon oil?

EXERCISE NO. 3

Object : To prepare Compound orange spirit

Requirements : Terpeneless lemon oil, terpeneless orange oil, anise oil, corinader oil, ethanol (90%) etc.

Theory : It is an alcoholic solution of oils which permits uniform introduction of small proportion of oils and also preserves orange and lemon oils from rapid oxidation. These oils should be brought in small quantities by the pharmacist and the spirit is most satisfactorily made from oils taken from bottles not previously opened. This will insure that delicacy of flavor which should always be characteristic of elixirs.

Formula

Terpeneless lemon oil	1.3 ml
Terpeneless orange oil	2.5 ml
Anise oil	4.25 ml
Corinader oil	6.25 ml
Ethanol (90%), sufficient to produce	1000.0 ml

Procedure

1. Measure terpeneless lemon oil, terpeneless orange oil, anise oil and coriander oil accurately and mix with ethanol (90%).
2. Make up the volume using ethanol (90%).
3. Mix it by shaking and store in a well-closed container.

Category : Flavoring agent and carminative **Weight per ml :** 0.8280 to 0.8410 g.

Storage : Lemon spirit should be kept in a well-filled container and protected from light.

Uses : Flavoring agent and carminative.

VIVA VOCE QUESTIONS

1. What is orange spirit?
2. Why this spirit is known as compound orange spirit?
3. What is the role of coriander oil in this preparation/
4. Give storage condition of spirit.
5. What is the use of anise oil?

EXERCISE NO. 4

Object : To prepare soap spirit

Requirements : Soft soap, ethanol (90%), etc.

Formula

Soft soap	65.0 g
Ethanol (90%), sufficient to produce	100.0 ml

Procedure

1. Dissolve soft soap in 30.0 ml of the ethanol.
2. Add sufficient ethanol 90% to produce 100 ml.
3. Allow to stand for a few hours and decant.

Category : For external use **Weight per ml :** 0.9450 to 0.9700 g

Storage : Preserve in a well-closed container and store in cool place.

Uses : For external use

VIVA VOCE QUESTIONS

1. What is soft soap?
2. What is the use of soft soap?
3. Why spirits are preserves in cool place?

EXERCISE NO. 5

Object : To prepare Surgical spirit

Requirements : Methyl salicylate, diethyl phthalate, castor oil, industrial methylated spirit, etc.

Formula

Methyl salicylate	0.5 ml
Diethyl phthalate	2.0 ml
Castor oil	2.5 ml
Industrial methylated spirit, Sufficient to produce	100.0 ml

Procedure

1. Mix measured quantity of methyl salicylate, diethyl phthalate, castor oil, and industrial methylated spirit.
2. Add sufficient industrial methylated spirit to produce 100 ml and mix.

Category : Surgical dressing

Storage : Preserve in a well closed container and put label on the container as Caution : 'This preparation is inflammable. Keep away from a naked flame'.

Uses : Surgical dressing

VIVA VOCE QUESTIONS

1. What is surgical spirit.
2. How to prepare surgical dressing?
3. What is the use of methyl salicylate?
4. What is the use of castor oil?
5. What is the requirements for label?

EXERCISE NO. 6

Object : To prepare Aromatic spirit ammonia

Requirements : Nutmeg oil, Lemon oil, ammonium bicarbonate, strong ammonia solution, ethanol and purified water.

Synonyms : Spirit of Sal Volatile

Theory : Generally the spirits are prepared by solution method but aromatic ammonia spirit is prepared by distillation method, because the oil of Nutmeg contains non-volatile resinous matter due to which the preparation changes to dark color on storage. Different distillation parts are collected separately. First part of distillate contains a major part of alcohol, volatile oils and a small proportion of water. This part contains volatile oils hence can not be subjected to heat. The second portion of distillate contains a major part of water, a little quantity of alcohol and volatile oil. Second portion of distillate is heated with ammonium bicarbonate to converts it into ammonium carbonate, completely process is performed in a closed vessel to prevent the evaporation of carbon dioxide which is produced during heating. The conversion of ammonium bicarbonate to ammonium carbonate is necessary before mixing it with reverse distillate because ammonium bicarbonate is insoluble in alcohol. Freshly prepared ammonium carbonate is used because ammonium carbonate on storage partially changes into ammonium bicarbonate. Due to this reason the bottles are well filled and due to presence of volatile ingredients it is stored in a cool place well-closed container.

Formula

Nutmeg oil	0.3 ml
Lemon oil	0.5 ml
Ammonium bicarbonate	2.5 g
Strong ammonia solution	67.5 ml
Ethanol (90%)	75.0 ml
Purified water, sufficient to produce	100.0 ml

Procedure

1. Take a mixture of lemon oil, nutmeg oil, ethanol and 37.5 ml of purified water in a distillation assembly.
2. Collect 87.5 ml of first distillate in separate container.
3. Second distillate of about 5.5 ml collected in separate flask.

4. Add the ammonium bicarbonate and strong ammonia solution in second distillate by heating on a water-bath to 60° C in a sealed bottle of not less 50 ml capacity.
5. Shake the mixture occasionally until solution is formed.
6. Cool and filter through absorbent cotton.
7. Mix the filtrate with the first distillate.
8. Add sufficient purified water to produce 100.0 ml and mix properly.

Category : Stimulant

Dose : 1 to 4.0 ml

Storage : Preserve in a well-closed container and keep in cool place.

Uses : Stimulant

VIVA VOCE QUESTIONS

1. How to prepare aromatic ammonia spirit?
2. Why this spirit can not prepared by simple solution method?
3. Why first and second distillate are collected?
4. Why first distillate is not heated?
5. What is the use of spirit?
6. How spirits are stored?
7. What is the role of lemon oil in spirit?

EXERCISE NO. 7

Object : To prepare Peppermint spirit

Requirements : Peppermint oil, ethanol (90%), etc.

Theory : Peppermint oil is obtained by distillation from the fresh overground parts of the flowering plants of *Mentha piperita* Linn. It is a colorless or pale yellow liquid having a strong penetrating odor of peppermint and a pungent taste, followed by a sensation of cold when air is drawn into the mouth. The odor of fresh peppermint is due to the presence of about 2.0% of a volatile oil, much of which is lost due to drying of

leaves in air. It is used as a flowering agent, carminative, antiseptic, and local anaesthetic. It is also used as a flavoring agent in candy, chewing gum, etc.
Peppermint oil and alcohol are volatile in nature, therefore, the preparation is stored in a well-closed container and in a cool place.

Formula

Peppermint oil	10.0 ml
Ethanol (90%), sufficient to produce	100.0 ml

Procedure

1. Take measured quantity of peppermint oil and mix with ethanol.
2. If the solution is not clear, shake with purified talc (5%) and filter it.
3. Add sufficient amount of ethanol to make up the volume.

Category : Flavoring agent and carminative

Dose : Usually, 1 ml three times a day.

Storage : Store in a well-closed container and in a cool place.

Uses : Flavoring agent and carminative in flatulence and nausea.

VIVA VOCE QUESTIONS

1. What is the source of peppermint oil?
2. How will you obtain peppermint oil from herb?
3. How will you prepare spirit?
4. What is the role of talc in the preparation of spirit?
5. What is pharmaceutical application of peppermint spirit?
6. Give storage condition of spirit.

EXERCISE NO. 8

Object : To prepare Chloroform spirit

Requirements : Chloroform, Ethanol (90%), etc.

Theory : Chloroform is used as analgesic, anaesthetic and muscle relaxant. It is less as anaesthetic as some times it is poisonous due to formation of phosgene gas during the storage in the presence of light. Chloroform spirit is used as carminative, flavoring agent, sweetening agent and preservative. When it is exposed to light or air, oxidation takes place. If the alcohol is added in small proportion it retards the oxidation. It should be stored in a well-closed, completely filled container due to volatile nature as well as for retarding oxidation.

Formula

Chloroform	5.0 ml
Ethanol (90%), sufficient to produce	100.0 ml

Procedure

1. Mix chloroform with ethanol (99%) and shake it.
2. Add sufficient ethanol to produce 100.0 ml of spirit and mix

Category : Pharmaceutical aid

Dose : 0.3 to 1.0 ml

Storage : Preserve in a air tight container and store in cool place.

Uses : Sweetening agent, carminative and preservative

VIVA VOCE QUESTIONS

1. Why chloroform is stored in amber color bottle?
2. Why chloroform spirit is prepared?
3. Give application of chloroform.
4. What is carminative?
5. Give storage condition of spirit.

EXERCISE NO. 9

Object : To prepare Spirit of ether

Requirements : Anaesthetic ether, ethanol (90%), etc.

Theory : Anesthetic ether is also known as Diethyl-ether. It is highly inflammable, therefore, keep away from an open flame and stored in an air tight container. Ether when administered by inhalation produced anesthesia. When administered by oral route it gives narcotic action similar to that of alcohol. It has a carminative action and its spirit is used in various pharmaceutical preparations.

Formula

Anaesthetic ether	33.0 ml
Ethanol (90%), sufficient to produce	100.0 ml

Procedure

1. Mix measured quantity of ether in ethanol.
2. Add sufficient quantity of ethanol 90% to produce volume upto 100.0 ml.

Category : Pharmaceutical aid

Dose : 1 to 4 ml

Storage : Preserve in an air tight container and store in a cool place

Uses : Pharmaceutical aid, also used as carminative and narcotic.

VIVA VOCE QUESTIONS

1. What is ether?
2. How will you prepare spirit?
3. Define carminative, and give` examples of flavoring agent.
4. What is pharmaceutical application of spirit of ether?
5. What is anaesthetic ether?

17

Suppositories

- Bismuth subgallate suppositoryies
- Glycerol suppositories
- Glycerated gelatin suppositories
- Tannic acid suppositories

A suppository is a solid or semi-solid mass meant to be inserted into a body orifice rectum, vagina and to lesser extent in the urethra to provide either a local or a systemic therapeutic effect. Suppositories can provide systemic effect when used rectally. They are frequently used for local effects for relief of hemorrhoids or infection in the rectum, the vagina or the urethra.

CLASSIFICATION

Suppositories are mainly of three types

1. Rectal suppositories : Rectal suppositories for adults are tapered at one or both ends and usually weighing about 2.0 g each. Infant suppositories usually weigh about one-half that of adult suppositories. Sedatives, tranquilizers, and analgesics are administered by rectal suppositories. They are used as hemorrhoid remedies dispensed over-the-counter. Adult suppositories of 2.0 g weight is based on use of cocoa butter as the base. If other bases are employed in place of cocoa butter, weight may be increased or decreased.
2. Vaginal suppositories : They are usually globular, or oviform and weighing about 5.0 g each. Vaginal medications are available in various physical forms like cream, liquids, gels for the management of vaginal disorders.
3. Urethral suppositories : They are not described in USP. They are cylindrical dosage form, diameter 5.0 mm, and length 50 mm for female and 125 mm for male. Weight

for urethral male suppository is 4.0 g while for female, weight is 2.0 g. It is not popular dosage form and is rarely encountered.

SUPPOSITORY BASES

Preparation of Suppositories

1. Hand Mould Suppositories : This is the oldest and simplest method of preparing of suppositories. A skill person is required for the preparation of suppositories. General process for preparation of suppositories is as follows :

    a. Mix measured quantity of medicinal substances with sufficient quantity of theobroma oil.
    b. Triturate and if required soften with diluted alcohol and rub until a smooth paste is formed.
    c. Add remaining quantity of theobroma oil and add wool fat for consistency.
    d. When the mass becomes plastic by vigrous kneading of the pestle quickly remove from the mortar with a spatula.
    e. Transfer with spatula to a piece of filter paper and keep in hands during the kneading and rolling procedure.
    f. Roll the mass by quick rotating movements of the hands and immediately place on a pile tile.
    g. A cylindrical suppository is formed by rolling the mass on the tile with a flat board.
    h. Cut in pieces by spatula.
    i. Give the shape by rolling one end on the tile with a spatula.
    j. Pack in butter paper or in proper container and store in cool place.

2. Compression mould suppositories (Cold compression)

    a. Mix theobroma oil and drug.
    b. Mixture is forced into a mould under pressure, using a wheel-operated press.
    c. Mould is removed, opened and replaced.
    d. On large-scale cold-compression machine are hydraulically operated by water-jacketed cooling and screw fed.

3. Fusion or melt mould suppositories

    a. Drug is dispensed or dissolved in a melted suppository base.
    b. Pour the mixture into suppository moulds allowed to cool in ice bath.
    c. Finish suppositories are remove by opening the mould.

d. Various types and sizes of moulds are available for preparation of suppositories. Moulds are made of aluminium alloys, brass or plastic and are available with from six to several hundred cavities.

4. Automatic mould Machine

All filling, ejection, and mould cleaning operation are fully automatic. The output of a typical rotary machine ranges from 3500 to 6000 suppositories per hours. The suppository mould is lubricated by brushing or spraying and then filled to a slight excess. Excess material is removed after the mass gets solidified and collected for re-use. All heating and cooling systems are fully automatic.

Suppositories are usually formulated on weight basis. The medicament replaces the portion of medicament having same density as that of theobroma oil. If the drug substance is heavier, it will replace a proportionally smaller amount of theobroma oil.

Example 1

Prepare six suppositories of 0.1 g tannic acid in each suppositories.

1. Density of tannic acid is 1.6 as compared with cocoa butter.
2. Weight of blank suppositories = 2.0 g
3. Cocoa butter replaced by drug = 0.1/1.6 = 0.062 g
4. Cocoa butter required for one suppository = 2.000 – 0.062 = 1.938 g
5. Actual weight of one suppositories = 1.938 + 0.10 = 2.038 g
6. Weight of cocoa butter for 8 suppositories (take excess weigh of 2 suppositories as required) = 1.938 × 8 = 15.504 g
7. Weight of tannic acid for 8 suppositories (take excess weigh of 2 suppositories as required) = 0.10 × 8 = 0.80 g
8. Total weight = 15.504 + 0.80 = 16.304 g

Example 2

Prepare 0.1 g tannic acid 6 suppositories using polyethylene glycol as base. Density factor of cocoa butter is 1.6 and density factor of polyethylene glycol is 1.25.

1. Weight of polyethylene glycol suppository = 1.75
2. Cocoa butter replaced by drug $= \frac{0.1}{1.6} \times 1.25 = 0.078$ g

3. Polyethylene glycol required for one suppository = 1.750 – 0.078 = 1.672 g
4. Actual weight of one suppositories = 1.672 + 0.10 = 1.772 g
5. Weight of polyethylene glycol for 8 suppositories (take excess weigh of 2 suppositories as required) = 1.672 × 8 = 13.376 g
6. Weight of tannic acid for 8 suppositories (take excess weigh of 2 suppositories as required) = 0.10 × 8 = 0.80 g
7. Total weight = 13.376 + 0.80 = 14.176 g

Displacement value of solid medicaments

The quantity of medicament which displaces one part of cocoa butter is called displacement value. The volume of particular suppository mould is uniform but fill weight will vary according to the density of medicaments as compared with cocoa butter with which the mould was calibrated.

Determination of Displacement value

Weight of six suppositories of theobroma oil or any other base = a g
Weight of six suppositories containing, say 40 percent of drug(s) = b g
Calculate weight of theobroma oil = 60 / 100 × b = c g
Calculate drug(s) = 40/100 × b = d g
(a – c) g = the weight of theobroma oil displaced by d g of drug substances
Displacement value of the medicament = d / (a – c)

Displacement Value(DV) of Solid Medicaments

Substances	DV	Substances	DV	Substances	DV
Alum	2.0	Hydrocortisone	1.5	Phenol	1.0
Aminophyllin	1.5	Ichthamol	1.0	Quinine HCl	1.0
Bismuth oxynitrate	5.0	Iodoform	4.0	Resorcinol	1.5
Bismuth subgallate	2.5	Lead acetate	3.0	Salol	1.5
Boric acid	1.5	Lead Iodide	5.0	Silver Proteinate	1.5
Chloral hydrate	1.5	Mercury ointment	1.5	Tannic acid	1.0
Cinchocaine	1.5	Morphine HCl	1.5	Zinc oxide	5.0
Cocaine HCl	1.5	Opium	1.5	Zinc sulphate	2.0
Hydrocortisone	1.5	Peptone	1.5		

Suppository Bases

1. Cocoa butter or theobroma oil
2. Water soluble or dispersible bases
3. Glycerinated gelatin

Quality Control Parameters

1. Appearance
2. Weight
3. Disintegration
4. Melting behavior
5. Mechanical strength
6. Content of active ingredients
7. Release pattern

PACKAGE AND STORAGE

They are packaged in partitioned boxes or in tightly closed screw-capped glass container or wrapped individually in aluminium foil. In large scale production suppositories are wrapped individual or separate in foil packs. Candy wrapping is the more familiar aluminium foil or PVC polyethylene strip. Low melting point suppositories should be store in cool place and theobroma oil suppositories should be store in refrigeration.

Mould Lubricant

Mould lubricants are used to remove theobroma suppositories without damaging their surfaces unless a lubricant is used. Lubrication is essential for glycero-gelatin bases because of their sticky nature. Lubricant selected on the base because it will be absorbed and fail to provide a buffer film between the suppositories and the metal. Following lubricant is used for the preparation of theobroma oil suppositories.

Soft soap	10.0 g
Glycerin	10.0 g
Ethanol, (90%)	50.0 ml

Lubricant must be compatible with medicament or adjuncts. In industry silicone fluid is used as lubricant. Mould is lubricated using a pad of guaze or muslin or with a small fairly stiff brush. Cotton wool is not used because some fibers adhere to the mould. Excess of lubricant can be removed by inverting the mould on a clean white tile.

EXERCISE NO. 1

Object : To prepare Compound bismuth subgallate suppositories

Requirements : Bismuth subgallate, resorcinol, zinc oxide, castor oil, theobroma oil.

Formula

Bismuth subgallate	200.0 mg
Resorcinol	60.0 mg
Zinc oxide	120.0 mg
Castor oil	60.0 mg
Theobroma oil, sufficient to fill a mould	q.s.

Procedure

1. Melt theobroma oil and add other additives.
2. Add remaining amount of theobroma oil and mix uniformly.
3. Properly lubricate the suppository mould.
4. Pour the hot base in the moulds and cool immediately in ice bath.
5. Excess of substance is scrapped off and re-used in next lot.
6. Open the mould and remove the suppository carefully.
7. Pack in butter paper and store in a well-closed container.

Category : Astringent and antacid **Dose :** 1 g.

Storage : Compound bismuth subgallate suppositories should be stored at a temperature not exceeding 20°C.

Uses : Astringent and antacid

EXERCISE NO. 2

Object : To prepare Glycerol suppositories

Requirements : Gelatin, glycerol, purified water, etc.

Formula

Gelatin	14.0 g
Glycerol	70.0 g
Purified water, sufficient to produce	qs

Procedure

1. Soak gelatin in purified water for about five minutes or until thoroughly softened.
2. Drain well.
3. Add glycerin and heat on water bath.
4. Evaporate excess quantity until the mixture weighs 100 g.
5. Pour the products into suitable moulds.

Category : Rectal evacuant

Storage : Preserve in a well-closed container, store in cool place.

Uses : Rectal evacuant

EXERCISE NO. 3

Object : To prepare Glycerin suppositories

Requirements : Glycerin, sodium stearate, purified water, etc.

Formula

Glycerin	91.0 g
Sodium stearate	9.00 g
Purified water	5.00 g

Procedure

1. Heat the weighed amount of glycerin in a suitable container to about 120°C.
2. Sodium stearate is dissolved in the heated glycerin with stirring.
3. Add purified water and mixed.
4. Pour the hot mixture in the suitable mould.

Category : Local application

Storage : Preserve in a suitable container and protected from the atmospheric moisture and temperature.

Uses : Local application

EXERCISE NO. 4

Object : To prepare Glycerated gelatin suppositories

Requirements : Drug, gelatin, glycerin, etc.

Formula :

Drug and purified water	10.0 g
Gelatin	20.0 g
Glycerin	70.0 g

Procedure :

1. Dissolve the drug in purified water.
2. Moistened the gelatin.
3. Heat the glycerin and add gelatin and drug solution.
4. Mix uniformly and this mixture poured in a mould.

Category : vaginal suppositories for local application of antimicrobial agent.

Storage : Preserve in a suitable container and protect from the atmospheric moisture and temperature
Uses : Antimicrobial

EXERCISE NO. 5

Object : To prepare Tannic acid suppositories

Requirements : Tannic acid, theobroma oil, etc

Formula

Tannic acid	0.2 g
Theobroma oil, sufficient to produce	qs

Procedure

1. Add lubricant in the mould and remove the excess of lubricant by draining.
2. Cool the lubricated mould over ice.
3. Theobroma oil heated in a porcelain dish over a water bath and avoid over heating.
4. Remove dish from the water bath as soon as the mass is molten
5. Weighed amount of tannic acid is powdered.
6. Add the powdered tannic acid in the molten theobroma oil.
7. Mix the tannic acid by spatula (sometimes molten theobroma oil divided in two parts).
8. Sufficiently thickened and flows sluggishly and pour the mixture into the holes in the moulds.
9. Filling each mould and stirring the contents of the dishes during the pouring.
10. Place the mould on ice and remove excess mass carefully by sharp knife.
11. When the suppositories have completely set hard, open the mould and remove the suppositories by slight pressure on the broad ends.
12. Remove any lubricant on final product, by rolling on the filter papers.

Category : Astringent

Storage : Preserve tannic acid glycerin suppositories in a tight container.

Uses : Astringent

VIVA VOCE QUESTIONS

For Exercise No 1 to 5

1. What are suppositories?
2. How suppositories are classified?
3. What is theobroma oil?
4. Why cocoa butter suppositories are obsolete?
5. How mould lubricants are selected?
6. What are ideal characteristics of mould suppositories?
7. Which conditions are suitable for the storage of suppositories?
8. Suppositories are not very popular dosage form why?
9. Give advantages and disadvantages of suppository.
10. How to calculate displacement value of medicaments?
11. What is the mechanism of suppositories base?
12. Name medicaments are used in preparation of suppositories?

18

Suspensions

- Amoxycellin trihydrate suspension
- Suspension containing wetting and aggregating agent
- Suspension containing low solid content
- Antacid suspension

A pharmaceutical suspension is a two-phase system in which insoluble solids are dispersed in a liquid medium. The particles size of the dispersed solid is usually greater than 0.5 µm. An aqueous suspension is useful formulation system for administering an insoluble or poorly soluble drug. They are widely administered from various route.

1. Oral route : Mixture, oral suspension
2. External use : Lotion
3. Ophthalmic use : Eye drops
4. Parenteral use : Intramuscular injection

Purpose of Suspensions

1. Suspension has large surface area of dispersed drug ensuring a high availability for dissolution and fast absorption.
2. Aqueous suspension is suitable for ophthalmic and parenteral administration of drug.
3. Suspension overcome stability problems of drugs that are unstable in solution form.
4. Suspension is suitable for drugs that are insoluble in water.
5. Suspension can overcome taste problems.
6. Suspensions are suitable for the preparation of sustained release dosage form.

Formulation of Suspension

An important decision should be taken before the preparation of suspension whether it is flocculated or deflocculated. Additives are selected on the basis of suspension. Following additives are commonly used in the preparation of suspension.

1. Flocculating agent
2. Suspending agent
3. Wetting agent
4. Dispersants
5. Preservatives
6. Organoleptic additives

Classification of Suspension

1. Flocculated suspension
2. Deflocculated suspension

Relative Properties of Suspensions

S.No.	Flocculated suspension	Deflocculated suspension
1.	Particles form loose aggregation	Particles exist in suspension as separate entities
2.	Rate of sedimentation is high	Sedimentation rate is slow
3.	A sediment is formed rapidly and re-disperse easily by shaking	A sediment is formed slowly and difficult to re-disperse by shaking
4.	Particles do not bond tightly to each other and a hard dense cake does not form	Repulsive forces between particles are overcome and a hard cake is formed
5.	The suspension is somewhat unsightly due to rapid sedimentation and the presence of an obvious, clear supernatant region.	The suspension has a pleasing appearance because suspended particles remains suspended for long time

Evaluation of Suspension

1. Sedimentation rate
2. Viscosity
3. Particle size
4. pH
5. Organoleptic properties

Problem in Suspension

1. Aggregation of particles
2. Cake formation
3. Locking of cap
4. Changes of Organoleptic properties
5. Crystal growth
6. Change of pH
7. Redispersibility

EXERCISE NO. 1

Object : To prepare Amoxycellin trihydrate suspension

Requirements : Amoxicillin trihydrate, sodium carboxymethyl cellulose, color tartrazine, sodium benzoate, sugar, pineapple flavor, purified water, etc.

Formula

Amoxicillin trihydrate	4.0 g
Sodium CMC	1.8 g
Color tartrazine	0.02 g
Sodium benzoate	0.45 g
Sugar	90.0 g
Pineapple flavor	1.0 g
Purified water, sufficient to produce	100.0 ml

Procedure

1. Dissolve separately color tartrazine, sodium benzoate, and sugar.
2. Mix and add sodium CMC and triturate in pestle mortar.
3. Disperse amoxycillin trihydrate and add flavor.
4. Add sufficient amount of water to produce 100 ml.

Category : Antibiotic

Dose : Orally for adult and children weighing 20 Kg or more, 250 to 500 mg every 8 h.

Storage : Preserve in a well-closed container. **Uses :** Antibacterial

EXERCISE NO. 2

Object : To prepare Suspension containing wetting and aggregating agent

Requirements : Cortisone acetate, Polysorbate 80, sodium CMC, sodium chloride, benzyl alcohol, purified water, etc.

Formula

Cortisone acetate*	2.5%
Polysorbate 80	0.4%
Sodium CMC	0.5%
Sodium chloride	0.9%
Benzyl alcohol	0.9%
purified water, sufficient to produce	100.0 ml

Procedure

1. Dissolve all additives in purified water except cortisone acetate.
2. Dispersed cortisone acetate.
3. Add sufficient purified water to produce 100 ml.
4. Pass this dispersion through a colloidal mill.

Category : Adrenocortical hormone

Dose : 50 to 300 mg by mouth

Storage : Preserve in a well-closed container, protect from light.

Uses : Adrenocortical hormone

---

* Other drug use in place of cortisone acetate

Note : Suspending agent, isotonic, wetting agent, and preservatives can be change of the same potency or concentration can also vary as per requirement of the suspension.

EXERCISE NO. 3

Object : To prepare Suspension containing low solid content

Requirements : Steroid, Polysorbate 80, sodium citrate, benzyl alcohol etc.

Formula

Steroid*	2.5%
Polysorbate 80	0.2%
Sodium citrate	1.0%
Benzyl alcohol	0.9%
purified water, sufficient to produce	100.0 ml

Procedure

1. Dissolve all additives in purified water except steroid.
2. Dispersed steroid.
3. Add sufficient purified water to produce 100 ml.
4. Passed this dispersion through a colloidal mill.

Category : Hormone

Storage : Preserve in a well-closed container, protect from light.

Uses : Adrenocortical hormone

EXERCISE NO. 4

Object : To prepare Antacid suspension

Requirements : Aluminium hydroxide gel, sorbitol, methylparaben, propylparaben, saccharin,peppermint oil, alcohol, purified water, etc.

---

* Other drug may be use in place of steroid

Note : Suspending agent, isotonic, wetting agent, and preservatives can be changed and potency or concentration may vary as per requirement.

Formula

Aluminium hydroxide gel	36.0%
Sorbitol	7.0%
Methylparaben	0.20%
Propyl paraben, saccharin	0.20%
Peppermint oil	0.005%
Alcohol	1.00%
Purified water, sufficient to produce	q.s.

Procedure

1. Dissolve methyl paraben, propyl paraben, saccharin, peppermint oil in alcohol and purified water.
2. Disperse aluminium hydroxide gel and mix uniformly.
3. Add sorbitol or mannitol and study the effect of concentration of these substance and their concentration in the suspension.

Category : Antacid

Dose : 7.5 to 15.0 ml

Storage : Preserve in a well-closed container and avoid freezing

Uses : Antacid

VIVA VOCE QUESTIONS

1. What are biphasic system?
2. What is difference between suspension and emulsion?
3. How will you prepare suspension?
4. What is the size of suspended particles?
5. What is antacid?
6. Which material is not suitable in the form of suspension?
7. What is the dose of suspension?
8. Why suspension is required?
9. Give pharmaceutical application of the suspensions.
10. Give storage and labeling condition of suspension.

19

Syrups

- Simple syrup
- Lemon syrup
- Orange syrup
- Vasaka syrup
- Tolu syrup
- Ginger syrup
- Cocoa syrup
- Invert syrup
- Codeine phosphate syrup

Syrups are concentrated oral solutions of sugar or nearly saturated solutions of sucrose in water or other aqueous liquids. 85% w/v or 65% w/w syrups will retard the growth of microorganisms. It is important to note that sucrose concentration should not reach the saturation point as a saturated solution of the sucrose may lead to crystallization of a small amount of sucrose due to change of physical condition viz. temperature. Dilute solution of sucrose provides an excellent nutritional source for yeast, moulds and other microorganisms. When heat is employed for the preparation of syrups, it small portion changes in dextrose and levulose. This phenomenon is called as inversion. Sucrose solution is optically active & rotates polarized light to right while on heat optical activity decreases, rotate the light to left due to formation of other compounds (dextrose ad levulose). The rate of inversion is enhanced by the presence of acids and hydrogen ions, which acts as a catalyst.

$$\underset{\text{sucrose}}{C_{12}H_{22}O_{11}} + H_2O = \underset{\text{dextrose}}{C_6H_{12}O_6} + \underset{\text{levulose}}{C_6H_{12}O_6}$$

Syrups are mainly of three types

Simple syrup : It contains sucrose in purified water alone or in combination of other polyols such as glycerin or sorbitol. These substances are added in syrups to reduce the crystallization of sucrose or enhance the solubility of excipients. (e.g. simple syrup etc.)

Medicated syrups : It contains some added medicinal substances in the syrups and used for therapeutic purpose (e.g. ephedrine sulphate syrup, chlorphineramine maleate syrup)

Flavored syrup : It contains various aromatic or pleasantly flavored substances but are non-medicated and generally used as vehicle or as a flavoring agent or for preservation. (e.g. cherry syrups, tolu balsam syrup, cocoa syrup etc.)

METHOD OF PREPARATION

Preparation of the syrup depends on the physical and chemical characteristics of the substance employed for its preparation. Four methods are commonly used for preparation of syrups.

1. Agitation without heat : This method is used for the preparation of syrup containing volatile substances. In this process active substance is added in solution and agitated in a glass-stoppered bottle. Closing of bottle is necessary to protect the syrup form contamination and loss of solution during the process. For large quantities, glass lined tank with mechanical agitators is employed for the preparation of syrups. This method is used for the preparation of wide variety of syrups. Cough syrups are commonly prepared by this process. (e.g. guainefenesin syrup, codeine syrup, ephedrine sulfate syrup etc.)

2. Solution with heat : This process is generally preferred as it is simple & less time consuming method , particularly if the constituents are not effected by heat and have non volatile in nature. In this process sucrose is added in the aqueous solution and heated till the sucrose is dissolved completely. Adding remaining amount of distilled water makes up volume of the solution. If the syrups containing any protogeneous substances are coagulated, it can be separated subsequently by straining. The concentration of the syrup is measured using saccharometer at warm condition if the specific gravity of the solution is known. Excessive heating of syrup is not suitable because more inversion of sucrose occurs with the increase in temperature. Syrups can not be sterilized in autoclave without carmelization. This solution is converted in yellowish to brown color due to formation of caramel by the effect of heat on sucrose.

3. Addition of a medicated liquid : This method is put to use in those cases in which tinctures, fluid extracts or other medicated substances in liquid forms are added to syrup to medicate it. In this process some time precipitation takes place due to presence of resinous and oily substances. It is necessary to take care that medicated substance should not get precipitate in this process.

4. Percolation : In this process, purified water or an aqueous solution is allowed to pass study through a bed of crystalline sucrose. A pledget of cotton is put in the neck of the percolator and purified water or aqueous solution is added in the percolator containing sucrose. The flow rate is controlled by the stopcock and maintained such that drops appear in rapid succession. If required, a small portion of liquid is re-passed through the perculator to dissolve the sugar completely in the liquid or aqueous solvent.

PRESERVATIVES

USP suggests that syrup should be kept at low temperature, about 25°C. Concentration without supersaturation is suitable for preservation. Following preservatives are used to prevent bacterial and mould growth viz. methylparaben, sodium benzoate, benzoic acid, glycerin etc. In combination of alkyl esters of p-hydroxybenzoic acid are effective to inhibit the yeast growth in syrups.

LABELLING

Label of the syrup should have following informations such as :

1. Date of manufacturing
2. Expiry date
3. Storage conditions

STORAGE

Syrups should be kept in well-closed containers and stored at temperature below 30°C. Bottle should be completely filled, carefully stoppered and stored in cool dark place.

Note : Syrup should be recently prepared or used within few months unless it contains appropriate anti-microbial agents or preservatives.

EXERCISE NO. 1

Object : To prepare Simple syrup

Requirements : Sucrose, distilled water, percolator, heat source, etc.

Theory : Syrups are concentrated nearly saturated solution of sucrose in purified water. It may or may not contain medicinal substances. It is used as a vehicle due to following reasons.

(a) It has sweet taste and act as a good flavor.
(b) Partially hydrolysis of sucrose into reducing sugar such as dextrose and levulose which act antioxidants.
(c) Nearly saturated solution of sucrose has great osmotic pressure. Water from the microorganism is withdrawn when syrup comes in contact and destruction of micro-organism occurs due to dehydration.

Fill the cool syrup in the bottle at room temperature or if hot syrup is filled in the bottle it should be occasionally shaked until it reaches to room temperature otherwise condensate water molecule may dilute the upper layer of the syrup which causes fermentation. Syrup is stored in loosely stoppered containers because carbon dioxide is evolved if fermentation take place which increases the pressure inside the bottle and bursting of container may occur. Moisture content inside the bottle is the cause of fermentation hence it should be always stored in dry bottle. Warm and cool place is not suitable for storage as it causes fermentation and crystallization.

Formula

Sucrose		66.7 g
Purified water,	sufficient to produce	100.0 g

Or

Sucrose		85.0 g
Purified water,	sufficient to produce	100.0 ml

Procedure

1. Transfer the weighed amount sucrose in a conical flask.
2. Add sufficient amount of water.
3. Make the syrup by heating or without heat.
4. Macerate it with cold water or heat the solution. It can also be agitated to accelerate the solubalization.

5. If necessary filter the syrup.
6. Store in a dried container with proper label.

Category : Pharmaceutical aid (sweetening agent or sweet vehicle)

Specific gravity : Specific gravity of syrup is not less than 1.30

Storage : Syrup should be store in a loosely stoppered bottle at fairly cool temperature. 0.15% methylparaben may be used as a preservative.

Uses : Pharmaceutical aid as sweetening agents, sweet vehicle and used as the base for various flavored and medicinal syrups.

VIVA VOCE QUESTIONS

1. What is syrup?
2. Why you should store syrup, at fairly cool temperature?
3. What is fermentation?
4. How to reduce fermentation of syrup?
5. Give factors accelerating in fermentation.
6. Why syrup is to be preserved in a loosely stoppered bottle?
7. Give pharmaceutical application of syrups.
8. How syrup act as a self preservative?
9. What is the concentration of sucrose suitable for preparation of syrup?
10. Give name and concentration of preservatives generally used in syrup.

EXERCISE NO. 2

Object : To prepare Lemon syrup

Synonyms : Citric acid syrup

Requirements : Lemon oil, citric acid monohydrate, invert syrup, simple syrup, ethyl alcohol, purified water

Formula

Lemon spirit	0.5 ml
Citric acid monohydrate	2.5 g
Invert syrup	10.0 ml
Syrup, sufficient to produce	100.0 ml

Lemon spirit (A)

Lemon oil	100.0 ml
Ethyl alcohol (96%) sufficient to produce	100.0 ml

Procedure

1. Cut thin slices of lemon and prepare lemon spirit by maceration process using 60 percent alcohol.
2. Dissolve the citric acid monohydrate in simple syrup.
3. Add freshly prepared invert syrup and lemon spirit, prepared using formula A & B.
4. Mix it properly.
5. Add sufficient amount of syrup to produce 1000 ml.
6. Lemon syrup has a weight per ml of about 1.33 g

Category : Pharmaceutical aid

Storage : Store in a closed container at a temperature not exceeding 25°C

Uses : Pleasant vehicle

VIVA VOCE QUESTIONS

1. What is lemon?
2. What is the use of lemon syrup?
3. How to prepare lemon spirit?
4. What is the alcoholic content in the syrup?
5. Give pharmaceutical application of lemon syrup.
6. How will you prepare invert syrup?

Note : lemon syrup should not be used if has a terebinthine odor or taste or have any other types of indications of deterioration.

EXERCISE NO. 3

Object : To prepare Orange syrup

Requirements : Orange peel, ethyl alcohol, sucrose, purified water

Theory : Fresh orange peel is obtained from outer part of the pericap of the ripe or nearly ripe fruit of *citrus chrysocarpa* or *citrus aurantium.* It is used for the prepration of the orange tincture. The outer part of the bitter orange peel contains a volatile oil and glycoside hesperidin. Hesperidin is hydrolysed by sulphuric acid and gives hesperitin, rhamnose and D-glucose. It has been used as bitter. Sweet orange peel has no inner portion of the rind. It contains volatile oil and no hesperidin because the glycoside is present in the white portion of the rind. It is used as a flavor.

Formula :

Orange tincture		6.0 g
Syrup	sufficient to produce	100.0 g

Orange tincture

Fresh orange peel		85.0 g
Ethyl alcohol	sufficient to produce	100.0 ml

Procedure

1. Prepare orange tincture by the maceration process.
2. Mix the orange tincture with syrup and make up the volume.

Category : Pharmaceutical aid

Storage : Store in a loosely stoppered bottle at fairly cool temperature.

Uses : Sweetening agent and flavored agent

VIVA VOCE QUESTIONS

1. What is orange syrup?
2. What is the use of orange syrup?
3. Give pharmaceutical importance of orange syrup.
4. How to prepare orange tincture?
5. What is the difference between orange tincture and orange syrup?

EXERCISE NO. 4

Object : To prepare Vasaka syrup

Requirements : Vasaka, ethyl alcohol, glycerin, syrup, distilled water

Theory : Vasaka (Arusha) consists of the fresh and dried leaves of *adhatoda vasica* (Family- Acanthaceae). Dried leaves are dull brown and light greyish brown colour. It is used as a expectorant.

Formula

Vasaka liquid extract	50.0 ml
Glycerin	10.0 ml
Syrup sufficient to produce	100.0 ml

Vasaka liquid extract

Vasaka powder (sieve no. 40)	100.0 g
Ethyl alcohol (40%) sufficient to produce	100.0 g

Procedure

1. Prepare vasaka liquid extract by percolation process.
2. Mix vasaka liquid extract with measure amount of glycerin and shake thoroughly.
3. Add sufficient quantity of syrup and mix properly.
4. Store in a suitable container and put label properly.

Category : Expectorant

Dose : 2 to 4 ml

Storage : In a well-closed container

Uses : Expectorant

VIVA VOCE QUESTIONS

1. What is the source of vasaka?
2. What is the use of vasaka syrup?
3. Give pharmaceutical application of vasaka syrup.
4. Give storage condition of the vasaka syrup.
5. What is the dose of vasaka syrup?

EXERCISE NO. 5

Object : To prepare Tolu syrup

Synonym : Balsam of Tolu syrup

Requirements : Tolu balsam, sucrose, distilled water, etc.

Theory : Tolu balsam is a solid or semi-solid balsam obtained by incision from the trunk of *Myroxylon balsamum* (Family-Leguminosae). It contains about to 80% resin, 7% volatile oil, 12-15% of free cinnamic acid, 2-8% of benzoic acid and 0.05% of vanillin. Tolu balsum is boiled with purified water to dissolve resinous matter contining cinnamic acid and benzoic acid. It is necessary to heat the tolu balsam because these substance are highly soluble in hot water is enhanced. Tolu balsum is boiled in covered dish so that enhance the solubility of tolu due to large surface area and reduces loss of water and aromatic volatile constituents of the balsam. The preparation contains 66.0% w/w of sucrose which acts as preservative due to higher osmotic pressure.

Formula

Tolu balsam	1.25 g
Sucrose	66.0 g
Purified water, sufficient to produce	100.0 g

Procedure

1. Boil 400 ml of purified water in dish or tared vessel.
2. Add weighed amount of tolu balsam in boiled water.
3. Cover the vessel partially and boil the contents for 30 min. with frequently stirring.
4. Add purified water to make the contents of the vessel about to 360 g.
5. Cool it and filter the solution.
6. Add sucrose in the solution and warm it on the water bath to dissolve sucrose completely.
7. Add sufficient purified water to make 1000 g of the solution.

Category : Expectorant

Dose : 2 to 8 ml (weight per ml is 1.310 to 1.330 g at 20°C)

Storage : Store in a loosely stoppered container at fairly cool temperature.

Uses : Expectorant

VIVA VOCE QUESTIONS

1. What is the source of tolu balsam?
2. Why is to be boiled tolu balsam?
3. Why the container is to be covered partially when tolu balsam heated ?
4. What is the use of sucrose in tolu syrup?
5. Give pharmaceutical application of the tolu balsam syrup.
6. Give name of chemical constituents in tolu balsam which acts as a expectorant.
7. What is the category of the tolu balsam syrup?
8. What is expectorant? Give other examples of syrups used as expectorant?
9. Give method of preparation of tolu balsam syrup.
10. What is the use of tolu balsam syrup?

EXERCISE NO. 6

Object : To prepare Ginger syrup

Requirements : Ginger powder, sucrose, ethyl alcohol, distilled water

Theory : Ginger is the rhizome of *Zingiber officinale* (Family-zingiberaceae), scraped to remove the dark outer skin and dried in the sun light. It has agreeble pungent taste and aromatic odor. It contains pungent substance gingerol, volatile oil and terpenes d-camphene, zingiberine, citral cineol and borneol. It is used as a flavoring agent, intestinal stimulant and carminative in colic and in diarrhea.

Formula

Strong ginger tincture	5.0 ml
Syrup sufficient to produce	100.0 ml

Strong Ginger Tincture

Ginger in moderately coarse powder	50.0 g
Ethyl alcohol (90%) sufficient to produce	100.0 ml

Procedure

1. Prepare strong ginger tincture by percolation process.
2. Add syrup and mix it thoroughly.
3. Add sufficient amount of syrup and make the volume.
4. Store in a closed container and put a proper label.

Category : Carminative

Dose : 2 to 4 ml (weight per ml is 1.290 to 1.310)

Storage : Preserve in a well closed container.

Uses : Carminative, flavoring agent, and as an intestinal stimulant .

VIVA VOCE QUESTIONS

1. What is the source of ginger?
2. Which part of ginger is used in medicine?
3. What is the use of ginger syrup?
4. What do you mean by term carminative?
5. What is the use of ethyl alcohol in syrup?
6. Give advantage of ginger syrup over other carminative syrup.

EXERCISE NO. 7

Object : To prepare Codeine phosphate syrup

Synonym : Codeine syrup

Requirements : Codeine phosphate, chloroform, ethylalcohol, sucrose, and distilled water

Theory : Codeine phosphate ($C_{18}H_{21}O_3N,H_3PO_4,\frac{1}{2}H_2O$) is the phosphate of the alkaloid codeine with half molecule of water. It is fine white needle-shaped crystals or hexagonal prisms or white crystalline powder. It is freely soluble in purified water and slightly soluble in alcohol. It is used as analgesic and antitussive. In this preparation main medicament is codeine phosphate which is used for the relief of cough and pain. Chloroform spirit acts as carminative, flavoring agent and preservative action. Simple syrup acts as a sweetening agent. This preparation contain volatile substance hence need to be stored in a well closed container and protected from light and air as chloroform is oxidized and gives toxic substances harmful to health.

Formula

Codeine phosphate	0.5 g
Chloroform spirit	2.5 ml
Purified water	1.5 ml
Syrup sufficient to produce	100.0 ml

Procedure

1. Weighed amount of codeine phosphate is to be dissolved in purified water.
2. Add chloroform spirit and mix thoroughly.
3. Add 75 ml of syrup and mix it.
4. Add sufficient amount of syrup to make 100 ml.
5. Kept in a well-closed narrow mouth amber colored bottle with suitable label.

Category : Analgesic and antitussive

Dose : 2 to 8 ml (each ml of codeine phosphate contains 5 ml of codeine sulphate)

Storage : Preserve codeine phosphate in a well closed container and protected from light and air.

Uses : Analgesic and antitussive (weight per ml of codeine phosphate is 1.30 at 25°C)

VIVA VOCE QUESTIONS

1. What is codeine phosphate?
2. Give pharmaceutical application of codeine phosphate syrup.
3. Why you prepare codeine phosphate syrup?
4. What is the use of the chloroform in syrup?
5. Why syrup is stored in closed bottle?
6. Give storage condition of syrup.
7. What is the dose of codeine phosphate?
8. Give advantages and disadvantages of this preparation.

EXERCISE NO. 8

Object : To prepare Acacia syrup

Requirements : Acacia powder, sodium benzoate, vanilla tincture, and sucrose

Formula

Acacia (granular or powdered form)	10.0 g
Sodium benzoate	0.1 g
Vanilla tincture	0.5 ml
Sucrose	80.0 g
Purified water, sufficient to produce	100.0 ml

Procedure

1. Mix the acacia powder, sodium benzoate and sucrose.
2. Add 42.5 ml of purified water and mix thoroughly.
3. Heat the mixture on steam bath until solubalised.
4. Cool the solution and remove the scum.
5. Mix vanilla tincture and add sufficient amount of purified water to make 100 ml.
6. Make homogeneous syrup and fill in suitable container.

Category : Pharmaceutical aid

Storage : Preserve in a well closed bottle.

Uses : Flavoring agent and demulcent.

VIVA VOCE QUESTIONS

1. What is the use of syrup of acacia?
2. What is acacia?
3. What is the role of sucrose in acacia syrup?
4. Give pharmaceutical application of acacia syrup.
5. Give method of preparation of the syrup.

EXERCISE NO. 9

Object : To prepare Cocoa syrup

Synonym : Cocoa syrup, chocolate syrup

Requirements : Cocoa, sucrose, liquid glucose, glycerin, vanillin, sodium benzoate, etc.

Formula

Cocoa	18.00 g
Sucrose	60.00 g
Liquid glucose	18.00 g
Sodium benzoate	0.10 g
Glycerin	5.00 ml
Sodium chloride	0.20 g
Vanillin	0.02 ml
Purified water, sufficient to produce	100.00 ml

Procedure

1. Mix sucrose and cocoa properly.
2. Add solution of liquid glucose, glycerin, sodium chloride, vanillin and sodium benzoate in 325 ml of purified water.
3. Heat the solution and boil it for about 5 min.
4. Cool the solution at room temperature.
5. Add sufficient quantity of purified water and make volume up to 100 ml.
6. Mix uniformly and preserve in a suitable container.

Category : Pharmaceutical aid

Storage : Store in a well closed container

Uses : Flavoring agent and should not be used for allergic patient

VIVA VOCE QUESTIONS

1. What is the use of cocoa syrup?
2. Give method of preparation of the syrup.
3. Give application of cocoa syrup.
4. What is cocoa?
5. What is the use of sodium benzoate in the syrup?

EXERCISE NO. 10

Object : To prepare and submit Invert syrup

Requirements : Sucrose, hydrochloric acid, and distilled water

Theory : The hydrolysis of sucrose by boiling with mineral acid or by the enzyme invertase produces a mixture of equal molecule of D-glucose and D-fructose. Sucrose solution is dextrorotatory but after hydrolysis it becomes laevorotatory.

$$C_{12}H_{22}O_{11} + H_2O \xrightarrow[\text{or invertase}]{H^+} C_6H_{12}O_6 + C_6H_{12}O_6$$

The specific rotation of sucrose is + 66.5°. D-glucose has specific rotation +52° and D-fructose has specific rotation -92°. The net specific rotation is -

$$+52° - 92°/2 = -20$$

In hydrolysis the specific rotation changes from + 66.5° to – 20°. i.e. (+) to (–) or said to be invert. Hence the hydrolysis of sucrose to D-glucose and D-fructose is termed as inversion and the hydroysis mixture is called as invert sugar. The relative sweetness of sugars is sucrose (100), fructose (173), dextrose (74), and lactose (16).

Formula

Sucrose	66.7 g
Purified water, sufficient to produce	100.0 g
Hydrochloric acid	qs
Sodium carbonate solution as neutralizing agent	

Procedure

1. Prepare syrup of sucrose 66.7% w/w in purified water.
2. Add hydrochloric acid slowly with continuous stirring.
3. Neutralizing the solution using sodium carbonate solution.
4. Identified inversion of simple syrup to invert syrup by measuring optical activity of syrup.
5. Take 1.0 g of syrup and dilute with 10 ml of distilled water and add 5 ml of potassium cupri-tartrate solution, a red precipitate is formed.

Category : Pharmaceutical aid

Storage : Invert syrup should be stored at 35° to 45°C because fructose is thermolabile and syrup darkens if not stored in cool place.

Uses : Sweetening agent, flavoring agent and used as a vehicle.

VIVA VOCE QUESTIONS

1. What is inversion?
2. What is invert sugar?
3. How sucrose is converted in invert sugar?
4. Give pharmaceutical application of invert syrup.
5. How to identify formulation of invert syrup?
6. What is the difference between invert syrup and honey?
7. How to control the reaction involved in inversion?
8. Why syrup is stored at below 45°C.

20

Tablets

- Phenobarbital tablets
- Chewable laxative tablets
- Ferrous sulphate tablets
- Aminophylline tablets
- Acetylsalicylic acid tablets
- Soluble acetylsalicylic acid tablets
- Effervescent aspirin tablets
- Paediatric dispersible aspirin tablets
- Dispersible aspirin and codeine tablets
- Compound magnesium trisilicate tablets
- Acetominophen tablets
- Vitamin B_1 tablets
- Chewable antacid tablets

Tablet is a solid unit dosage form and is prepared by compression using tablet compression machine. It can be defined as solid, flat or biconcave disc (also available in various shapes) prepared by compressing a drug or a mixture of drugs with or without suitable diluents. About 90% of drugs are available in the form of tablets to produce a therapeutic effect when administered by the oral route. e.g. Aspirin tablet, paracetamol tablet, ketoprofen tablets, flurbiprofen, nimesulide, chloramphenicaol, albendazole, APC, Cyproheptadine, warfarin tabletes etc.

Advantages

Tablet dosage form have number of potential advantages over the other solid dosage form as well as liquid dosage forms such as.

1. Unit dosage form (dosage accuracy)
2. Compactness of dosage
3. Easy to swallow
4. Highly soluble
5. Flexibility of dosage form
6. Easy to prepare
7. Economic for packaging, shipping and storage

Disadvantages

1. Highly amorphous substances are very difficult to compress.
2. Poor wetting and slow dissolution drugs can not be placed in the form of tablets.
3. Objectionable odor, bitter taste and humectant substances need special treatment for compression.

FORMULATIONS

Formulation of tablets requires following additives.

1. Diluents : Diluents are substance which increases bulk and convert in the compressible form, when drug material is potent or in adequate to provide a suitable shape and size to tablet (for ingestion size range of tablets are 3/16 to 1/2 inch, and if size is less than 3/16 inch it becomes difficult to handle and if size is more than 1/2 inch it is difficult to swallow. Tablets weigh ranges from 120 to 700 mg for spherical tablets and in case of oval tablets it is upto 800 mg or more. e.g. Dibasic calcium phosphate, calcium sulphate, lactose, lactose anhydrous, lactose spray dried, mannitol, sorbitol, sucrose, dextrose. Following are direct compressible diluents, Sta-Rx-1500, Emdex (contain dextrose 90 to 92% and maltose 3 to 5%), Celutab (dextrose & maltose), microcrystalline cellulose, etc.

2. Binders and adhesive : These materials are used in dry or liquid form to reduce the amorphous nature of substance and convert into compressible form (wet granulation). e.g. Acacia, tragacanth, gelatin, alginates, methylcellulose, hydroxypropylmethylcellulose, hydroxypropylcellulose, PVP, Starch, sorbiol, ethylcellulose, Pregel-atinized starch, glucose, iris moss, ghatti gum, arabogalactan, waxes, water alcohol.
3. Disintegrants : Most of the tablet contains disintegrating agent. It facilitate the disintegration of the tablet in small particles when it contracts with the GIT contents. Breaking of tablet is based on the swellability, adsorption of water or chemical reaction. e.g. Soluble starch, pregelatinised starch (PGS), veegum HV, bentonite, microcrystalline cellulose, sodiumcarboxymethylcellulose, PVP, guar gum, Ispaghula, primogel, explotab, aerosol, natural sponge citrus pulp.
4. Glidants : Glidants act as a flow promoter and reduces the friction between particles. It improves the flow properties of granules or powder through hopper to die. It is not deformed by pressure of the tablet machine. e.g. Talc, starch, magnesium stearate, calcium stearate, boric acid, sugar, lycopodium and sodium chloride.
5. Lubricants : Lubricants reduce inter-particular friction. It improves the ejection of tablet from die wall and reduces the sticking problems and produce smooth tablets.

e.g. Talc, magnesium stearate, calcium stearate, stearic acid, polyethyleneglycols, starch derivatives.

6. Antiadhesives or antisticking agent : These materials are used to reduce the adhesion of the tablet surface to dies and punches during the compression of tablets. The pressure of the machine deforms these materials. It reduces sticking, picking and chipping problems. e.g. Paraffin, stearic acid, cocoa butter, soaps, stearic acids, starch, derivatives.
7. Coloring agent : Coloring materials or dyes are used in tablet formulation mainly for three purposes, disguising of off color drugs, product identification and production of more elegant products. e.g. water soluble dyes and many other FD&C approved colors or dyes.
8. Flavoring agent : These substances are not necessary for the formulation of compressed tablets. Special tablets require flavoring agents such as chewable tablet, lozenges, etc. e.g. flavoring oils like cinnamon, coriander, and caraway etc.
9. Sweeteners : Chewable tablets have sweetening agents because it is not swallowed. e.g. saccharin sodium, aspartame, sugar, etc.

Compressed tablets

For making compressed tablets, it is necessary to prepare the material in a dry, granular form to render it suitable for passing through a compressing machine. Following general processes are employed for this purpose.

I. Wet granulation

1. The drug or a mixture of drugs in powder form is mixed. Add a suitable inert substance to acts as diluent and half quantity of disintegrating agent. The material in the requisite degree of fineness is mixed and damped with a suitable moistening agent selected with regard to its effect on the chemical and physical nature of the material.
2. The moistened material is made into granules by passing it through a sieve no. 12, 16, 22 etc.
3. The granules are dried in a current of air, at a suitable temperature generally not exceeding 60°,
4. Pass through suitable sieve to get uniform granules, rejecting oversize and undersize, add half quantity of disintegrant, lubricant, antiadhesive, and glidant etc.
5. Mix these granules well using the spatula, suitable for compression to get the desired type of tablets.

II. Dry granulation

1. The drug or a mixture of drugs in powder form is mixed. If necessary, add a suitable inert substance to acts as diluent along with half quantity of disintegrating agent and lubricant.
2. Compress the powder materials in the form of large tablets.
3. These large tablets are broken up into granules of a suitable size.
4. Pass these granules through set of sieves to get uniform granules.
5. Add remaining quantity of disintegrant and lubricant and glidant. Mix properly using spatula. The granules can be compressed using tablet machine.

III. Direct compression

1. The drug or a mixture of drugs in powder form is mixed. If necessary, add a suitable inert substance to acts as diluent direct compressible, disintegrating agent, glidant and lubricant.
2. Pass the powder material through a suitable sieve.
3. Compress the powder materials in the form of tablets.

Stages involved in various granulation methods

Wet granulation	Dry granulation	Direct compression
Raw material (drug)	Raw material (drug)	Raw material (drug)
Weighing and measuring	Weighing and measuring	Weighing and measuring
Screening	Screening	Screening
Manual feeding	Blending (slow-speed planetary mixer)	Manual feeding
		-
Blending (slow-speed planetary mixer)	Compress slug	-
		-
Wetting (hand addition)	-	-
Subdivision (comminutor)	Subdivision (comminutor)	-
Drying (fluid bed dryer)	-	-
Subdivision (comminutor)	-	-
Premixing (Barrel roller)	Premixing (Barrel roller)	-
Lubrication (ribbon blinder)	Lubrication (ribbon blinder)	-
Manual feeding	Manual feeding	-
Compression	Compression	Compression

CLASSIFICATION OF TABLETS

Tablets are classified on the basis of shape and size, process involved in its preparation and route and other characteristics.

I. Shape and size

On the basis of shape and sizes tablets are classified as

1. Round tablets : a. flat surface b. concave surface c. convex surface tablets
2. Oval tablet : a. oral route b. other than oral route tablets
3. Oblong tablet : suppositories
4. Cylindrical tablet : suppositories
5. Triangular tablets : Lozenges and chewable tablets

II. Process

On the basis of process, tablets are divided in two parts

1. Compressed Uncoated tablet
2. Compressed coated tablets : a. sugar coated b. film coated c. enteric coated

III. *Route and other special characteristics*

Tablets are divided mainly in four category viz.

1. Oral tablets for ingestion

   (a) Compressed tablets (CT)
   (b) Multiple compressed tablets (MCT)
   (c) Prolonged action tablets
   (d) Sustained release tablets
   (e) Delayed action tablets
   (f) Chewable tablets
   (g) Coated tablets

2. Tablets used in oral cavity

   (a) Buccal tablets
   (b) Sublingual tablets
   (c) Troches and Lozenges
   (d) Dental cones

3. Tablets administered by other than oral route

   (a) Suppositories
   (b) Implant

4. Special tablets

   (a) Effervescent tablets
   (b) Dispersible tablets
   (c) Dispensing tablets
   (d) Hypodermic tablets
   (e) Tablet triturates (TT)
   (f) Soluble tablets

Evaluation Parameters

1. Shape and size : Thickness of the tablets should be controlled within ± 5% variation with a standard value. It may change with die-wall, particle size, distribution, packing of particles and compressive load. The crown thickness of tablets may be measured by micrometer (sliding caliper scale). This test is necessary for packaging of tablet as well as drug content uniformly.

2. Organoleptic properties : Some tablets may contain Organoleptic substances such as flavoring agent, coloring agent and sweetener. Color uniformity of the tablet can be evaluated by reflectance spectrophotometry, tristimulus colorimetry or microreflectance photometer for measurement of color uniformly.

3. Hardness : Suitable hardness is necessary for handling in manufacturing, packaging and shipping. Hardness of the tablets are measured by Monsanto tester, Strong-cobb-tester, Pfizer tester, Erweka tester and Schleuniger tester.

4. Friability : Friability is another measurement of tablet strength as tablet hardness is not an absolute indicator of strength. Roche friabilator is used as a laboratory equipment for the determination of friability. It has a plastic chamber that revolves at 25 rpm, tablet drops a distance of six inches with each revolution and operated for 100 revolutions. Conventional tablets granules lose weight less than 0.5 to 1.0%.

5. Weight variations : It is measured to ensure that tablet contains the proper amount of drug and is a satisfactory test for determination of content uniformity of tablets. Usually ten tablets are used for this test. In USP, 20 tablets are weighed individually and average weight of tablets is calculated.

USP limit		IP limit	
Average weight of tablets (mg)	Maximum variation	Average weight of tablets (mg)	Maximum variation
130 or less 130 to 324 More than 324	± 10.0% ± 7.5% ± 5.0%	80 or less 80 to 300 More than	± 10.0% ± 7.5% ± 5.0%

6. Disintegration : Breaking of tablets into smaller particles or granules is known as dis-integration and time taken for breaking of tablets in a suitable medium is called disintegration time (DT). It is determined by USP apparatus. It consists of 6 glass tube each 3 inches long, open at top and has 10 mesh screen at the bottom end of basket rack. One tablet is placed in each tube and placed in a one liter beaker of water, simulated gastric fluid or simulated intestinal fluid at 37 ± 2°C. It moves up and down through a distances of 5 to 6 cm. at 28 to 32 cpm. Uncoated tablet has disintegration times as low as 5 minutes. Majority of tablets has DT of 30 minutes. Disintegration time of enteric coated tablet is one hour in simulated gastric fluid and two hours in simulated intestinal fluid.

8. Dissolution : The rate of absorption depends on the dissolution of the product. It is directly proportional to the bioavailability of the products. For dissolution study two types of apparatus are used USP XX/NF XV, basket assembly and paddle assembly. It contains 1000 ml of hemispherical flask temperature is maintained at 37 ± 0.5°C, $t_{90\%}$ obtained within 30 min. is satisfactory.

8. Drug content : Drug uniformity in tablet is determined by assay. It is calculated batch to batch or lot to lot. It is estimated by the titration, spectrophotometer and HPLC.

TABLET DEFECTS

1. Capping : Partial or complete separation of top or bottom surface of the tablet from the main body is called capping. This type of problem is generally found after comp-ression.

   Causes : a. Entrapment of air b. deformational problems c. Incorrect set up of press.

   Remedies : a. Reduce the speed of press b. Increase percentage of moisture

2. Lamination : Separation of a tablet into two or three layers is called as lamination.

   Causes : a. Entrapment of air b. deformational properties of formulations

c. Direct compression d. Low moisture content less than 5%.
Remedies : a. Reduce the speed of press b. Increase percentage of moisture.

3. Picking : Removal of surface material from the punch surface is called picking
   Causes : a. Embossing or engraving of punch tips b. Convex type of punch
   c. Insufficient anti-adhesive.
   Remedies : a. Reduce convex curve of punch b.Uniform distribution of anti-adhesive c. Reduce letters like A,O, Welcome, etc. on punch.
4. Sticking : Some material of tablet is adhere inside the die wall, is called as sticking
   Causes : a. Insufficient lubricant b. Excessive moisture content.
   Remedies : a. Mix sufficient lubricant
5. Chipping : Serious problem of sticking is called chipping. The tablet produced has rough surface due to this problem. It creates difficulty in free movement of punches.
   Causes : a. Low melting point substances
   Remedies : a. Add higher melting point substances e.g. stearic acid.
6. Mottling : Unequal distribution of color in tablet is called mottling.
   Causes : a. Migration of dye b. Unequal distribution of color c. large particle size
   Remedies : a. Added anti-leaching dye substances in the product such as acacia, tragacanth.
7. Poor flow : Incomplete filling of dies by the powder material due to poor flow of material from hopper to die wall.
   Causes : a. Speed of tablet machine b. Different density of powder material as well as additives.
   Remedies : a. Addition of glidant e.g. talcum, colloidal silica.
8. Double Impression : One dark and one lighter impression on tablet surface is called double impression.
   Causes : a. Uncontrolled movement of lower punch.
   Remedies : a. Regulation of the punch movement as well as speed of the machine.

EXERCISE NO. 1

Object : To prepare Phenobarbital tablets

Requirements : Phenobarbital, lactose, starch, talc, mineral oils, etc

Formula : For each tablet

Phenobarbital	65.0 mg
Lactose, fine powder	40.0 mg
Starch	10.0 mg
Starch paste	qs
Talc	10.0 mg
Mineral oils	4.0 mg

Procedure : Wet granulation method

1. Mix Phenobarbital, lactose, and half quantity of dry starch.
2. Add sufficient starch paste (10%) to get damp mass.
3. Granulate by passing through a mesh-14 screen.
4. Dry the granules at 60°C.
5. Use 20-mesh screen to get uniform granules.
6. Add remaining amount of dry starch and talc, mix well.
7. Add the mineral oil and mix again.
8. Compress the granules using 9/32 inch die cavity punching machine.

Category : Anti-convulsant

Dose : 1 to 3 tablets

Storage : Preserve in a well-closed container.

Uses : Anti-convulsant

EXERCISE NO. 2

Object : To prepare Chewable laxative tablets

Requirements : Phenolphthalein, sugar, cocoa, calcium stearate, talc, gelatin, etc

Formula : For each tablet

Phenolphthalein	64.0 mg
Sugar, in powder	750.0 mg
Cocoa, in powder	350.0 mg
Calcium stearate	12.0 mg
Talc	60.0 mg
Gelatin (10% solution)	qs

Procedure : Wet granulation method

1. Mix the phenolphthalein, sugar, and cocoa, mix well.
2. Moisten the powder with gelatin solution.
3. Granulate by passing through a 8-mesh screen.
4. Dry the granules at 40°C to 50°C.
5. Dry granules passed through a 16-mesh screen.
6. Mix the calcium stearate and talc (previously passED through 100-mesh screen).
7. Compress the powder using 5/8-in. die cavity using flat faced punching machine.

Category : Cathartic

Dose : 1 to 2 tablets for adults

Storage : Preserve in a well-closed container

Uses : Cathartic, laxative

EXERCISE NO. 3

Object : To prepare Ferrous sulphate tablets

Requirements : Ferrous sulphate, corn starch, explotab, talc, magnesium stearate, sugar solution, etc.

Formula : For each tablet

Ferrous sulphate, dried	300.0 mg
Corn starch	60.0 mg
Explotab	4.0 mg
Talc	30.0mg
Magnesium stearate	4.0 mg
Sugar solution (20%)	qs

Procedure : Wet granulation method

1. Mix the ferrous sulphate, and half quantity of corn starch.
2. Moisten the powder with sugar syrup.
3. Granulate by passing through a 12-mesh screen.
4. Dry the granules overnight at 60°C to 70°C.
5. Dried granules are passed through a 18-mesh screen.
6. Add explotab, talc and magnesium stearate.
7. Compress the powder using 3/8 inch die cavity by deep cup punching machine.

Category : Haematinic

Dose : 1 to 2 tablets

Storage : Preserve in a well-closed container.

Uses : Haematinic

EXERCISE NO. 4

Object : To prepare Aminophylline tablets

Requirements : Aminophylline, tricalcium phosphate, pregelatinised starch, talc, mineral oil, purified water, etc

Formula : For each tablet

Aminophylline	100.0 mg
Tricalcium phosphate	50.0 mg
Pregelatinised starch	15.0 mg
Talc	30.0 mg
Mineral oil	2.00 mg
Purified water	qs

Procedure : Wet granulation method

1. Mix the aminophylline, tricalcium phosphate, and half quantity of pre-gelatinized starch.
2. Moisten the powder with water.
3. Granulate by passing through a 12-mesh screen.
4. Dry the granules overnight at 35°C to 40°C.
5. Dried granules passed through a 20-mesh screen.
6. Add talc and mineral oil, mix about 10 min.
7. Compress the powder using 5/16 inch die cavity by deep cup for enteric coating.

Category : Smooth muscle relaxant

Dose : 1 to 3 tablets

Storage : Preserve in a tightly closed container, protected from light

Uses : Smooth muscle relaxant

EXERCISE NO. 5

Object : To prepare Acetylsalicylic acid tablets

Requirements : Acetylsalicylic acid, lactose, starch, starch paste, talc, and stearic acid.

Formula : For each tablet

Acetylsalicylic acid	300.0 mg
Lactose	150.0 mg
Starch dry	20.0 mg
Starch paste	qs
Talc	15.0 mg
Stearic acid	15.0 mg

Procedure : Wet granulation method

1. Mix the Acetylsalicylic acid, lactose, half quantity of dry starch.
2. Moisten the powder with starch paste.
3. Granulate by passing through a 12-mesh screen.
4. Dry the granules for overnight at 65°C.
5. Dry granules are passed through a 20-mesh screen.
6. Add talc, dry starch and stearic acid.
7. Compress the powder.

Category : Analgesic **Dose :** 1 to 4 tablets

Storage : Preserve in a well-closed container **Uses :** Analgesic and acute rheumatism

EXERCISE NO. 6

Object : To prepare Soluble acetylsalicylic acid tablets

Requirements : Acetylsalicylic acid, citric acid, calcium carbonate, saccharin sodium.

Formula : For each tablet

Acetylsalicylic acid	300.0 mg
Citric acid	30.0 mg
Calcium carbonate	100 mg
Saccharin sodium	3 mg

Procedure : Prepared by Wet granulation or Dry granulation methods.

Disintegration : Maximum time three minutes **Category :** Analgesic

Dose : 1 to 3 tablets **Storage :** Preserve in a well-closed container.

Uses : Analgesic, and acute rheumatism

EXERCISE NO. 7

Object : To prepare Acetylsalicylic acid (Aspirin) tablets.

Requirements : Aspirin, starch, Cab-O-Sil, etc

Formula : For each tablet

Aspirin, pass through 20-mesh screen	325.0 mg
Starch, dried	32.5 mg
Cab-O-Sil	0.1 mg

Procedure : Dry granulation method

1. Mix the aspirin, dry starch, and Cab-O-Sil in a blender for 10 min.
2. Compress into slugs using 1-in. flat-face punches.
3. Reduce the slugs to small granules.
4. Pass the granules through a 16-mesh screen.
5. Transfer the granules to a tablet machine hopper.
6. Compress into tablets using 13/32 in. standard concave punches.

Disintegration : Maximum time three minutes **Category :** Analgesic

Dose : 1 to 3 tablets **Storage :** Preserve in a well-closed container.

Uses : Analgesic, and acute rheumatism

Note : All operations should be carried out in a dehumidified area at a relative humidity of less than 30% at 70°F during handling of aspirin

EXERCISE NO. 8

Object : To prepare Effervescent aspirin tablets

Requirements : Aspirin, citric acid, sodium bicarbonate, fumeric acid, etc.

Formula : For each tablet

Aspirin, granular, pass through 20-mesh screen	325.0 mg
Citric acid, fine granular	520 mg
Sodium bicarbonate	2.050 g
Fumeric acid, fine granular	305.0 mg

Procedure : Dry granulation method

1. Mix weighed quantity of aspirin, citric acid, sodium bicarbonate, and fumeric acid in a twin-shell blender mixer for 20 min.
2. Compress into slugs using 5/4-in. flat-face punches and compress slug to approximately 3/8-in. thick.
3. Grind the slugs to small granules.
4. Pass the granules through a 16-mesh screen.
5. Transfer the granules to a tablet machine hopper.
6. Compress into tablets using 7/8 in. flat-face beveled-edge punches.

Disintegration : Maximum time three minutes **Category :** Analgesic

Dose : 1 to 3 tablets **Storage :** Preserve in a well-closed container.

Uses : Analgesic, and acute rheumatism

EXERCISE NO. 9

Object : To prepare Calcium gluconate tablets

Requirements : Calcium gluconate, sucrose, mentha oil, etc.

Formula : For each tablet

Calcium gluconate	500.0 mg
Sucrose	1000.0 mg
Mentha oil	0.0015 ml

Procedure : Wet granulation method

1. Dissolve mentha oil in small amount of ethanol.
2. Prepare tablets by wet granulation.

Category : Electrolyte replenisher **Dose :** 2 to 5 tablets

Storage : Preserve in a well-closed container. **Uses :** Electrolyte replenisher

EXERCISE NO. 10

Object : To prepare Paediatric dispersible aspirin tablets.

Requirements : Aspirin, citric acid, calcium carbonate, saccharin sodium, etc.

Formula : For each tablets

Aspirin	75.0 mg
Citric acid	7.5 mg
Calcium carbonate	25.0 mg
Saccharin sodium	0.75 mg

Procedure : Prepared by dry granulation

Category : Analgesic

Dose : For children,
1 to 2 years : 1 to 2 tablets up to four times daily
3 to 5 years : 3 to 4 tablets up to four times daily
6 to 12 years : 4 tablets up to three or four times daily.

Storage : Paediatric dispersible aspirin tablets should be kept in a well-closed container and stored at a temperature not exceeding 25°C.

Uses : Analgesic

Note : Do not use more than two times, except under medical supervision.

EXERCISE NO. 11

Object : To prepare Dispersible aspirin and codeine tablets

Requirements : Aspirin, codeine phosphate, calcium carbonate, citric acid, saccharin sodium, etc.

Formula : For each tablet

Aspirin, in fine powder	400.0 mg
Codeine phosphate	8.0 mg
Calcium carbonate	130.0 mg
Citric acid	40.0 mg
Saccharin sodium	4.0 mg

Procedure : Prepared by dry granulation

Category : Analgesic

Dose : 1 or 2 tablets

Storage : Preserve in a well-closed container and protected from light. The label on the container states that the tablets contain 'Aspirin' unless this word appears in the name of the tablets. This requirements does not apply in countries where exclusive proprietary rights in the name aspirin are claimed.

Uses : Analgesic

EXERCISE NO. 12

Object : To prepare Compound magnesium trisilicate tablets

Requirements : Magnesium trisilicate, dried aluminium hydroxide gel, peppermint oil, etc.

Formula : For each tablets

Magnesium trisilicate	250.0 mg
Dried aluminium hydroxide gel	120.0 mg
Peppermint oil	0.003 ml

Procedure : Prepared by dry granulation

Category : Antacid **Dose :** 1 to 3 tablets

Storage : Preserve in a well-closed container. The label on the container states that the tablets should be chewed before being swallowed.

Uses : Antacid

EXERCISE NO. 13*

Object : To prepare Acetaminophen tablets

Requirements : Acetaminophen, avicel PH 101, stearic acid, etc

Theory : Smaller crystalline size of acetaminophen is desired to improve the dissolution. For this purpose higher proportion of avicel PH 102 is used in place of avicel PH 101 and to be used as a glidant. All lubricant should be screened before being added to blender.

Formula : For each tablet

Acetaminophen, granular or large crystal	325.0 mg
Avicel PH 101	138.35 mg
Stearic acid, fine powder	1.65 mg

Procedure : Direct compression method

1. Mix acetaminophen and avicel in blender.
2. Add screened stearic acid and mix.
3. Compress the tablets using 7/16-in. standard concave or flat beveled punches.

Category : Analgesic, antipyretic, anti-inflammatory **Dose :** 1 to 3 tablets

Storage : Preserve in a well-closed container.

Uses : Analgesic, antipyretic, anti-inflammatory

EXERCISE NO. 14

Object : To prepare Vitamin B_1 tablets

Requirements : Thiamine hydrochloride, avicel, lactose, magnesium stearate, Cab-O-Sil.

Formula : For each tablets

Thiamine hydrochloride	100.0 mg
Avicel PH-102	83.35 mg
Lactose anhydrous	141.65 mg
Magnesium stearate	6.65 mg
Cab-O-Sil	1.65 mg

Procedure : Direct compression method

1. Mix thiamine hydrochloride, avicel, lactose and Cab-O-Sil in blender.
2. Add screened magnesium stearate and mix for 5 min.
3. Compress the tablets using 13/32 inch using standard concave punches.

Category : Vitamin preparation

Dose : 1 to 3 tablets

Storage : Preserve in a well-closed container.

Uses : Vitamin

EXERCISE NO. 15

Object : To prepare Chewable antacid tablets

Requirements: Dried aluminium hydroxide gel, magnesium stearate, sucrose, mannitol, polyvinypyrrolidine, etc.

Formula : For each tablet

Aluminium hydroxide, dried gel	400.0 mg
Magnesium hydroxide, fine powder	80.0 mg
Sucrose	20.0 mg
Mannitol, fine powder	180.0 mg
Polyvinylpyrollidine (10%) solution	30.0 mg

Procedure : Wet granulation method

1. Mix dried aluminium hydroxide gel, magnesium stearate, sucrose, and Mannitol.
2. Moisten with polyvinylpyrollidine (10%) solution in ethanol.
3. Granulate by passing through a 14-mesh screen.
4. Dry the granules overnight at 45°C to 60°C.
5. Dry other additives and compress the granules using ½-in. flat-face beveled-edge punches.

Category : Antacid

Dose : 1 or 2 tablets, repeated in accordance with the needs of the patient.

Storage : Preserve in a well-closed container. The label on the container states that the tablets should be chewed before swelling. Tablets should be stored at a temperature not exceeding 25°C.

Uses : Antacid

VIVA VOCE QUESTIONS

For Exercise No 1 to 15

1. What are tablets?
2. Classify tablets with examples.
3. What are the advantages and disadvantages of tablets over the other dosage forms?
4. What is unit dosage form?
5. How antacid tablets are prepared?
6. Define glidant, lubricant and anti-adhesive.
7. What is the function of glidant in the tablet formulation?
8. Why diluent is used in the preparation of tablet?
9. How will you prepare tablets? Explain with examples?
10. How method of preparation of tablet is selected?
11. What is disintegrant?
12. Why half quantity of the disintegrant in the formulation of tablet by wet granulation method is affected before and remaining after the formulation of the granules?
13. Give mechanism of the tablet machines?
14. Classify tablet machines.
15. What are chewable tablets?
16. Explain term capping, lamination and sticking.
17. Define direct compressible substances.
18. What is the difference between soluble tablets and dispersible tablets?
19. Give evaluation parameters of tablets.
20. What is the disintegration time of dispersible tablets?
21. Why coating on tablet is required?
22. What is enteric coating?
23. What are the various steps in the sugar coating?
24. What is the role of film coating?
25. What are the additives used for sugar coating?

21

Tinctures

- Belladonna tincture
- Compound benzoin tincture
- Compound cardamom tincture
- Orange tincture
- Aromatic cardamom tincture

Tinctures are defined as alcoholic or hydroalcoholic solutions prepared from vegetable materials or from chemical substances. Tinctures are generally prepared by two methods.

1. Percolation : This method is most frequently used to extract the active ingredients in the preparations of tinctures and fluid extracts. It contains a narrow cone-shaped vessel open at both ends known as percolator. The solid substances are moistened with a suitable amount of specified menstruum and allow to stand for about three to four hours in a well-closed container after that the drug mass is packed into the percolator. Saturate the mass with menstruum. Additional menstruum is added to give a shallow layer above the mass and the mixture is allowed to macerate in the closed percolator for 24 hours. The outlet of the percolator is then opened and the liquid is allowed to drip slowly. If necessary, additional menstruum is added until the volume of the percolate collected is about three quarters of the required. The marc is then pressed and the expressed liquid is added to the percolate. Sufficient menstruum is mixed to obtaine the sufficient volume. This mixture of various portion of percolate is clarified by filtration or allowed to stand for a few hours and then decant the percolate. e.g. Belladonna tincture

2. Maceration : In this process the solid substances are placed in a stoppered container with the whole of the solvent. Allowed it to stand for a period of at least three days until the soluble matter of the substances is dissolved. It is frequently agitated. The mixture is then strained and collected in a container. The damp solid material (marc) is pressed and combine both the liquids. This mixture is classified by filteration or by decantation after few hours standing with agitation. e.g. Sweet orange peel tincture, compound benzoin tincture.

CONTAINER

Narrow mouthed screw-capped colored fluted bottle

STORAGE

Tincture should be kept in tightly closed containers protected from light and stored in a cool place.

EXERCISE NO. 1

Object : To prepare Belladonna tincture

Requirements : Belladonna Herb, ethanol (70%), etc.

Formula

Belladonna Herb	100 g
Ethanol (70%), in moderately coarse powder	q.s.

Procedure

1. Prepare the tincture by percolation.
2. Determine the proportion of alkaloids in the tincture by assay (It contains in 2.0 ml, 0.6 mg of alkaloids of belladonna, calculated ashyoscyamine).
3. Add sufficient ethanol to produce a sufficient strength of belladonna.
4. Keep aside for twenty four hours and filter.

Category : Parasympatholytic **Dose :** 0.6 to 2.0 ml

Storage : Preserve belladonna tincture in a air-tight container, protected from light.

Uses : Parasympatholytic

Note : Following tinctures are discarded after some times because active principles changes with time. e.g. Aconite tincture, Belladonna tincture, and hyoscyamus tincture

EXERCISE NO. 2

Object : To prepare Compound benzoin tincture

Synonyms : Friars' Balsam

Requirements : Aloes, tolu balsam, prepared storex, sumatra benzoin, ethanol, etc.

Formula

Aloes	2.0 g
Tolu balsam	2.5 g
Prepared storex	7.5 g
Benzoin, crushed	10.0 g
Ethanol (90%), sufficient to produce	100.0 ml

Procedure

1. Macerate the aloes, tolu balsam, prepared storax and benzoin crushed with the 80 ml of ethanol in a closed vessel for not less than tw0 days.
2. Shake the tincture occasionally and filter.
3. Pass ethanol through the filter and make up the volume.

Category : Protective **Weight per ml :** 0.880 to 0.910 g

Dose : External preparation

Storage : Preserve Compound benzoin tincture in a well-closed container.

Uses : Protective

EXERCISE NO. 3

Object : To prepare Compound cardamom tincture

Requirements : Cardamom seeds, caraway, cinnamon, amaranth, glycerin, ethanol (45%), etc.

Formula

Cardamom seeds, in moderately coarse powder	1.4 g
Caraway, in moderately coarse powder	1.4 g
Cinnamon in moderately coarse powder	2.8 g
Amaranth	0.50 g
Glycerin	5.0 ml
Ethanol (45%), sufficient to produce	100.0 ml

Procedure

1. Moisten the mixed powders with a sufficient quantity of alcohol and prepare about 90 ml of tincture by percolation.
2. Add glycerin and mix.
3. Add solution of amaranth.
4. Add sufficient ethanol to produce tincture 100 ml.
5. Filter the tincture and preserve in a suitable container.

Category : Carminative **Dose :** 2 to 4 ml

Storage : Preserve in a well-closed container. **Uses :** Carminative

EXERCISE NO. 4

Object : To prepare Orange tincture

Requirements : Fresh orange peel, alcohol, etc.

Formula

Fresh orange peel, in thin slices	25 g
Ethanol (90%), sufficient to produce	100.0 ml

Procedure

1. Prepare the tincture by maceration process.
2. Add sufficient Ethanol (90%) to produce 100 ml.

Category : Flavoring agent **Storage :** Preserve in a well-closed container

Uses : Flavoring agent

EXERCISE NO. 5

Object : To prepare Aromatic cardamom tincture

Requirements : Cardamom oil, caraway oil, cinnamon oil, clove oil, strong ginger tincture, ethanol (90%), etc.

Formula

Cardamom oil	0.3 ml
Cinnamon oil	1.0 ml
Caraway oil	1.0 ml
Clove oil	1.0 ml
Strong ginger tincture	6.0 ml
Ethanol (90%), sufficient to produce	100.0 ml

Procedure

1. Mix all oils with ethanol about to 80 ml.
2. Add strong ginger tincture and mix properly.
3. Add sufficient quantity of ethanol to produce 100 ml.

Category : Carminative **Dose :** 0.12 to 0.60 ml

Storage : Preserve in a well-closed container. **Uses :** Carminative

VIVA VOCE QUESTIONS

For exercise 1 to 5

1. What are tinctures?
2. What is the difference between tincture and extraction?
3. How to prepare tincture?
4. What is the use of cardamom tincture?
5. What is the use of orange tincture?
6. Give pharmaceutical application of tincture.
7. What is percolation?
8. Why tincture is required?
9. What is the difference between tincture and lotion?
10. How to preserve tinctures?

22

Miscellaneous

APPLICATIONS

Applications are liquid or semi-solid preparations similar to lotions containing one or more active ingredients intended for application to the skin. Usually, they are solution, suspension or emulsions. They may contain appropriate anti-microbial preservatives unless the active ingredients or vehicles having sufficient intrinsic antibacterial and antifungal activity. They may contain other additives such as stabilizers. Most of the official applications are used as anti-parasitic.

Examples

1. Benzyl benzoate application
2. Application of Dicophane (Chlorophenothane application)
3. Application of Gamma Benzene Hexachloride
4. Nystatin application
5. Application of salicylic acid and sulfur

STORAGE

Application should be stored in a narrow-mouthed screw-capped bottle. Semi-solid preparation which are applied with a brush, a colored fluted wide-mouthed bottle is more suitable.

LABELLING

The label should mention

1. The application is intended 'For External use only'.
2. Expiry date
3. Storage conditions under which the application should be stored.

EXERCISE NO. 1

Object : To prepare Benzyl Benzoate application

Requirements : Benzyl benzoate, emulsifying wax, purified water, etc.

Theory : Benzyl benzoate is used as a scabicide. It can cause burning and skin eruptions when applied on the sensitive skin. This preparation should not be allowed to come into contact with the eyes. Benzyl benzoate is an oily liquid preparation, containing 25 percent of benzyl benzoate and is emulsified with 2 percent of emulsifying wax. This preparation is insoluble in water. Emulsifying wax is used as an emulsifying agent to mix the benzyl benzoate with water. The consistency of the product is enhanced after some time. Hence, it should be packed in a wide-mouthed (amber fluted) jar to facilitate the application using brush. Place the appropriate label on the container with proper direction as mentioned on the label.

Formula

Benzyl benzoate	2.5 g
Emulsifying wax	0.2 g
Purified water sufficient to produce	10.0 ml

Procedure

1. Weigh the benzyl benzoate into an evaporating basin and warm on the water bath until the benzyl benzoate is liquefied.
2. Melt the emulsifying wax in separate dish and add benzyl benzoate.
3. Mix and allow the mixture to cool to upto 60°C.
4. Warm an excess of water to 60°C in another container.
5. Pour the mixture into sufficient warm purified water to produce 100.0 ml and stir continuously at a fixed interval of time.
6. Replace the closure and shake well until cool (Use an electric stirrer for shaking).

Emulsifying wax

Cetostearyl alcohol	900 g
Sodium lauryl sulphate	100 g
Purified water, sufficient to produce	40 ml

Procedure

1. Melt the cetostearyl alcohol and heat to 95°C.
2. Add required amount of the sodium lauryl sulphate and mix it properly.
3. Pour purified water and heat to 115°C with constant stirring until frothing ceases and the product is converted in translucent form.
4. Cool quickly and store in a closed container.

Category : Scabicide

Dose : External use only. This preparation must not be allowed to come into contact with the eyes.

Storage : Application should be stored in a narrow-mouthed screw-capped bottle.

VIVA VOCE QUESTIONS

1. What is application?
2. What is benzyl benzoate?
3. Why this application should not be put on a cold surface during stirring.
4. What is the role of emulsifying wax in this preparation?
5. Give storage condition of application.
6. Define parasitic disease and its symptom.
7. Give application of benzyl benzoate.

EXERCISE NO. 2

Object : To prepare Chlorophenothane application

Synonym : DDT application, Dicophane application, etc.

Requirements : Chlorophenothane, emulsifying wax, xylene and citronella oil, etc.

Theory : Chlorophenothane contains not less than 75.0 percent of 1,1,1-trichloro-2,2-di-(p-chlorophenyl)ethane and not less 9.5% of hydrolysable chlorine. It is practically insoluble in water and soluble in 40 to 60 parts of alcohol.

Formula

Chlorophenothane	2.0 g
Emulsifying wax	4.0 g
Xylene	15.0 ml
Citronella oil	0.5 ml
Purified water, sufficient to produce	100 ml

Procedure

1. Melt emulsifying wax at a low temperature.
2. Dissolve the chlorophenothane and citronella oil in xylene.
3. Mix the solution with emulsifying wax and shake it.
4. Warm purified water in a separate container to 60°C.
5. Pour the solution into purified water at same temperature and stir thoroughly.

Category : Arthropod toxicant

Storage : Preserve chlorophenothane in a well-closed container & protected from light.

VIVA VOCE QUESTIONS

1. What is DDT? Give IUPAC name of DDT.
2. Give storage condition of this application.
3. How to prepare chlorophenothane application?
4. What is chlorophenothane? Give application of chlorophenothane.

EXERCISE NO. 3

Object : To prepare Gamma Benzene Hexachloride application

Synonym : B.H.C. application

Requirements : Gamma Benzene Hexachloride, emulsifying wax, xylene, Citronella oil

Formula

Gamma Benzene Hexachloride	100.0 mg
Emulsifying wax	4.0 g
Xylene	15.0 ml
Citronella oil	0.5 ml
Purified water, sufficient to produce	100 ml

Procedure

1. Melt emulsifying wax at a low temperature.
2. Dissolve Gamma Benzene Hexachloride and citronella oil in the xylene.
3. Mix the solution with the emulsifying wax and shake it continuously.
4. Warm purified water in a separate container at 60°C.
5. Pour the solution into the purified water at a same temperature and stir thoroughly.

Direction for use : Rub about 15 ml with the fingers into hair and roots of the hair. Do not wash the head for the next twenty four hours. Protect the eyes.

Storage : Store in a well closed container.

VIVA VOCE QUESTIONS

1. What is BHC? Give application of BHC.
2. Give method of preparation of gamma benzene hexachloride application.
3. What is the role of xylene citronella oil in application?

DOUCHES

A douches is an aqueous solution directed against a part or into a cavity of the body. Douches are most frequently dispensed in the form of a powder with directions for dissolving in a specified quantity of warm water. If powder or tablets are employed for this purpose, they must be completely soluble for this purpose, they must be completely soluble in water to produce a clear solution.

Tablets are prepared by simple process but the lubricants and diluents must be soluble in water. For preparation of tablets, boric acid is used as lubricant and sodium chloride is used as a diluent.

Tablets deteriorate on exposure to moist air and should be stored in air tight conatainers. Douches are not official preparation. Some preparations contain alum, zinc sulphate, boric acid, phenol, sodium borate. Some compendia have reported a few substances such as benzylkonium chloride to be used in various douches and compound sodium borate solution in nasal or pharyngeal douche.

Douches are mainly used as

1. Antiseptic (e.g. chlorhexidine, chloroxylenol, lactic acid, mercurial chloride, potassium permagnate, etc.)
2. Astringent (e.g. alum, tannic acid, zinc sulphate, etc.)
3. Cleaning (e.g. sodium chloride, boric acid, saponated cresol, etc)
4. Sothing action

CLASSIFICATION OF DOUCHES

1. Eye douches : It is used to remove foreign particles and discharges from the eyes use allowed to run from the inner to the outer cornea of the eye.
2. Nasal douches : It is used to cleanse nose by using bulb syringes.
3. Pharyngeal douches : It is used in infection of throat for an operation.
4. Vaginal douches : It is used for irrigation of the vagina by using 8-10 oz capacity bulb syringes with a large vulcanite or rubber spray tube.
5. Rectum douches : It is used for irrigation of rectum using rectal syringe of the bulb type with a long narrow nozzle.

Container : It is dispensed in colored fluted bottles or single use plastic packs with rectal nozzles for enema.

Labelling : Comply with the general requirement of the label and additional information should be mentioned clearly on the label.

'For external use only'

'For nasal use Only' or 'For vaginal use Only' or 'For rectal use Only'

EXERCISE NO. 1

Object : To prepare Potassium permanganate douche

Requirements : Potassium permanganate, purified water, etc

Formula

Potassium permanganate	0.1 g
Purified water, sufficient to produce	100.0 ml

Procedure

1 Grind to weighed amount of potassium permanganate with purified water using pestle mortar.
2 Add more water and regrind.
3 Allow undissolved crystals to settle and pour the supernatant into a conical flask.
4 Filter through a clean sintered glass filter and make up the volume through filter.

Category : Antiseptic

Dose : 50 ml

Storage : Preserve in a suitable container.

Uses : Antiseptic

VIVA VOCE QUESTIONS

1. What are douches?
2. How will you prepare douches?
3. Name of the drugs used in douches?
4. Classify douches with examples.
5. How to apply douches?

EXERCISE NO. 2

Object : To prepare Astringent douche

Requirements : Alum, zinc sulphate, Alum, zinc sulphate, liquefied phenol, glycerin, purified water phenol, glycerin, purified water, etc.

Formula

Alum	0.4 g
Zinc sulphate	0.5 g
Liquefied phenol	0.6 ml
Glycerin	12.5 ml
Purified water, sufficient to produce	100.0 ml

Procedure

1. Grind weighed amount of alum with purified water.
2. Add zinc sulphate and dissolve.
3. Gradually add liquefied phenol and glycerin and shake.
4. Add sufficient purified water to produce 100 ml.
5. If necessary filter the douche.

Category : Astringent

Dose : 50 ml

Storage : Preserve in a well-closed container.

Uses : Astringent

VIVA VOCE QUESTIONS

1. What are douches?
2. What is the use of alum in douches?
3. What is alum?
4. What is the use of glycerin in douches?
5. What is the pharmacology of douches?
6. How will you store douches?

GARGLES

Gargles are aqueous solutions used for treating the pharynx and nasopharynx by forcing air from lungs through the gargle that is held in the throat. Gargles are generally dispensed in concentrated form. They must be diluted with water prior to its use. Gargles are pleasantly flavored and medicated than mouthwashes. Many mouthwashes are used as gargles, either as such or diluted with purified warm water. Gargles contain following substances viz.

1. Antibiotics (e.g. Benzocaine)
2. Antiseptics (e.g. Borax)
3. Anti-inflammatory (e.g. clove oil)
4. Anti-fungal (e.g. Phenol, thymol etc.)
5. Analgesic (e.g. Potassium permagnae)
6. Astringents (e.g. Benzylkonium chloride, potassium chlorate)
7. Alkalinizing agents (e.g. Cresote)
8. Deodorants (e.g. Sodium bicarbonate)
9. Local anesthetics (e.g. Eugenol)

Container : Narrow mouthed, screw capped, colorless fluted bottles.

EXERCISE NO. 1

Object : To prepare Phenol glycerin gargles

Requirements : Phenol glycerin (16% w/w phenol and 64% w/w glycerin) , amaranth solution (1.0% w/v in chloroform), purified water, etc.

Formula

Phenol glycerin	5.0 ml
Amaranth solution	1.0 ml
Purified water, sufficient to produce	100.0 ml

Procedure

1. Prepare phenol glycerin using phenol and glycerin.
2. Mix phenol glycerin with purified water and shake.
3. Add amaranth solution.
4. Add sufficient purified water to produce 100 ml.

Category : Local analgesic, local antiseptics

Storage : Preserve in a well-closed container.

Uses : Local analgesic, local antiseptics

VIVA VOCE QUESTIONS

1. What are gargles?
2. What is the difference between gargles and mouthwashes?
3. How will you prepare gargles?
4. What is the use of gargles?

EXERCISE NO. 2

Object : To prepare Potassium permanganate gargles

Requirements : Potassium permanganate, purified water, etc.

Formula

Potassium permanganate	25.0 mg
Purified water, sufficient to produce	100.0 ml

Procedure

1. Grind weighed amount of potassium permanganate with water.
2. Remove undissolved permanganate by filtration and add sufficient amount of purified water to produce 100 ml.

Category : Astingent **Storage :** Store in a well closed container.

Uses : Astringent

VIVA VOCE QUESTIONS

1. What are gargles?
2. What is the role of gargles in the new drug delivery system?
3. How many parameters are used to act as emulsifier?

Note : Phenol glycerin is diluted with glycerin to form the pharmaceutical preparation.

GLYCERITES

Glycerites are solutions or mixtures of medicinal substances in glycerin, containing not less than 50% by weight of glycerin. Most of glycerites are viscous and some are extensively used in the dispensing.

Glycerin is important pharmaceutical solvent forming permanent and concentrated solutions of most otherwise objectionable. Some of solutions are used in original form as medicinal agents while others are used to prepare aqueous and alcoholic dilutions of substances which are not readily soluble in water or alcohol.

Glycerites are sweet in taste and does not become rancid during storage. Certain aqueous or non-aqueous preparations are used to remove wax (cerumen) from the ear. Glycerin is hygroscopic in nature and should be stored in air tight closed container.
Following glycerites are generally dispensed in the laboratory for various purposes e.g.

1. Borax glycerin (e.g. bacteriostatic)
2. Starch glycerin (e.g. emollient)
3. Phenol glycerin (e.g. mouth ulcer and tonsilitis)
4. Tannic acid glycerin (e.g. in ulcer)
5. Ichthammol glycerin (e.g. lymphadentis and thrombophlebitis)

EXERCISE NO. 1

Object : To prepare Borax Glycerin

Requirements : Borax, glycerin, etc.

Formula

Borax	12.0 g
Glycerin	88.0 g

Procedure

1. Powder borax and pass through sieve 80.
2. Triturate fine powder of borax with glycerin.
3. Warm the solution with constant stirring until borax is dissolved.
4. Filter, if necessary.

Category : Bacteriostatic

Storage : Store in a well closed container and label as 'To be used sparingly'.

VIVA VOCE QUESTIONS

1. What are glycerites?
2. Why borax is used in this preparation?
3. What is the use of this preparation?
4. How it is used?
5. How will you store it?
6. Is there any reaction between borax and glycerin?
7. Why glycerin is used in such preparation?

EXERCISE NO. 2

Object : To prepare Boric acid glycerin

Requirements : Boric acid , Glycerin

Theory : Boric acid reacts with glycerin and form boroglycerin. The rate of formation of boroglycerin is directly proportional to heat but it should not be heated above 150°C. If heated above this temperature the glycerin gets converted into acrolein, which is toxic in nature and the color of the preparation is changed to brownish. The heating is done on a sand bath. This formulation is used as an antiseptic as ear drops and throat paint. This formulation is stored in a well-closed container as glycerin is hygroscopic in nature. Boric acid is used as local anti-infective. Glycerin is employed as a solvent of substances to be applied to the throat and mouth because its high viscosity prevents rapid dilution with saliva and consequently the local action of the drug is prolonged. It is used for this reason in glycerin of tannic acid, glycerin of boric acid, glycerin of borax and also used in the other preparations.

Formula

Boric acid	31.0 g
Glycerin, sufficient to produce	100.0 g

Procedure

1. Heat about half quantity of glycerin to a temperature of 140°C to 150°C in sand bath.
2. Add boric acid and heat continuously with stirring until it dissolves.

3. Evaporate at a temperature not exceeding 150°C until 20 percent of weight of the solution is reduced with continuous stirring.
4. Add sufficient warm glycerin to produce the required weight.

Category : Antiseptic **Storage :** Store in a well closed container.

VIVA VOCE QUESTIONS

1. How boric acid glycerin is prepared?
2. Give pharmaceutical application of boric acid.
3. What is acrolein?
4. What is the use of glycerin in this preparation?
5. Define hygroscopic substances.
6. Why preparation is stored in well-closed container?

EXERCISE NO. 3

Object : To prepare Tannic acid glycerin

Synonym : Tann. Acid Glycer.

Requirements : Tannic acid, sodium citrate, dried sodium sulphite, glycerin etc.

Theory : Tannic acid is a tannin usually obtained from nutgalls, the excrescences produced on the young twigs of *Quercus infectoria* Oliver and allied species of Quercus Linne (Family : Fagaceae). It is yellowish to light brown, glistening scales, or an amorphous powder. It is odorless or has faint characteristic odor and very soluble in water and in alcohol, practically insoluble in benzene, chloroform and in ether. Internally, tannic acid is used in the treatment of diarrhea, to precipitate bacterial toxins and supresses mucosal weeping.

Tannic acid preparations should not be used as the principal remedial agents because it produces excessive gastric irritation, nausea, and vomiting. 10% tannic acid solution was employed in the treatment of burns. This solution was sprayed over the affected area following careful debridement to form a firm ester that help to protect the burned tissue from infection, preserve body fluids and for comfort to the patient. It has limited application because tannic acid is not an active germicide. It is absorbed from denuded surfaces causing serious systemic toxicity to liver damage. It also causes necrosis of viable tissue in the burnt area. An ointment or glycerin of tannic acid has been employed in the treatment of minor burns, bed sores, weeping ulcers etc.

In this preparation sodium sulphite acts as anti-oxidant and preservative while sodium citrate acts as a chelating agent and forms a complex with iron present in glycerin as a impurity and prevents the reaction between tannic acid and glycerin. Specified temperature (between 115°C to 120°C) is essential for the preparation of tannic acid glycerin because above this temperature tannic acid gets converted into gallic acid, which is black in color.

Tannic acid is the main ingredient in this preparation, which acts as an astringent, and glycerin is used as a vehicle in this preparation. Astringents are substances, which causes contraction locally after topical application. It precipitates proteins when applied to mucous membrane or damaged skin and form a superficial protective layer. It also checks the exudative secretions and hemorrhage.

Formula

Tannic acid	20.0 g
Sodium citrate	1.0 g
Dried sodium sulphite	0.2 g
Glycerin	78.8 g

Procedure

1. Powder weighed quantity of tannic acid, sodium citrate and dried sodium sulphite in a porcelain dish with half quantity of glycerin until a smooth mixture is formed.
2. Add remaining quantity of glycerin in the mixture and mix well.
3. Heat the mixture with occasional stirring on a sand bath to a temperature between 115°C to 120°C until a homogeneous solution is obtained.
4. Store the solution in a closed container and fix appropriate label.

Category : Astringent

Dose : 'For External Use only'

Storage : Preserve tannic acid glycerin in a tight container,.

Uses : Astringent

VIVA VOCE QUESTIONS

1. What is tannic acid?
2. Why temperature should be below 120°C during preparation?
3. What is the use of glycerin, sodium citrate and other additives in this preparation?
4. Give pharmaceutical application of tannic acid glycerin.
5. Give use of anti-oxidant and chelating agents in this preparation.
6. What is the difference between sand bath and water bath?
7. Why sand-bath is used during heating in this preparation?

EXERCISE NO. 4

Object : To prepare Phenol glycerin

Requirements : Phenol, glycerin, etc

Theory : Phenol (C_6H_5OH) is a colorless, hygroscopic, crystalline solid. The liquid form of glycerin contains about 5% of water, known as carbolic acid. It is obtained from natural as well as synthetic source. Below 65.8°C phenol is only partially miscible with water but above this temperature it is miscible in all proportions. It is completely miscible in ethanol and other organic solvents. Phenol turns pink in color due to exposure of air and light. It has characteristic odor.

Phenol is a poison when taken orally and produces painful blisters in contact with skin. Phenol glycerin is too strong so it should be diluted with equal volume of glycerin to use as antiseptic and analgesic paint. This preparation should not be diluted with water because phenol is converted in phenoxy ions and becomes caustic in nature. It should be stored and prepared in dry container due to the same reason.

Phenol is generally used in proprietary antiseptic, mouthwashes, hemorrhoidal preparations and burn remedies. A few drop of liquefied phenol may be used to cauterize small wounds, dog bites, snakebites etc. Phenol is commonly employed as an antipruritic in form of lotion (1%), ointment (2%), or a simple aqueous solution (0.5 to 1.0%). Glycerin is used as a vehicle.

The preparation should be protected from atmospheric moisture because glycerin is hygroscopic in nature and the preparation will become caustic due to absorption of moisture.

Formula

Phenol	16.0 g
Glycerin	84.0 g

Procedure

1. Weighed quantity of phenol is mixed with glycerin and warm gently, if necessary, until solution is obtained.
2. Preserve phenol glycerin in a well-closed container with proper label.

Category : Local analgesic and antiseptic

Dose : 'For external use only'

Storage : Preserve in a well closed container

Uses : Local analgesic and antiseptic

VIVA VOCE QUESTIONS

1. What is phenol?
2. What is liquefied phenol?
3. Why glycerin is used as a vehicle in this preparation?
4. What will happen if this preparation is diluted with water?
5. Give source of phenol.
6. Give pharmaceutical application of phenol.
7. How phenol is stored?

Note : Aqueous solution of tannic acid is decomposed on keeping

EXERCISE NO. 5

Object : To prepare Starch glycerin

Requirements : Starch, benzoic acid, purified water and glycerin, etc.

Formula

Starch	10.0 g
Benzoic acid	0.2 g
Purified water	20.0 g
Glycerin	70.0 g

Procedure

1. Weighed quantity of starch and the benzoic acid should be triturated with the purified water in porcelain dish until a smooth mixture is produced
2. Add glycerin and mix well.
3. Heat the mixture on a sand bath to a temperature between 140°C and 144°C with constant but gentle stirring until a translucent, jelly like mass is formed, and then strain through muslin cloth.

Category : Emollient

Dose : For external use, topically as required.

Storage : Starch glycerin is preserved in a well closed container protected from light and moisture.

Uses : The preparation is used as a substitute for a fatty ointment and it has also been used as a pill excipient.

VIVA VOCE QUESTIONS

1. What is starch?
2. How to prepare starch glycerin?
3. Give storage condition of glycerin.
4. What is the use of benzoic acid in the preparation?
5. Give pharmaceutical application of starch glycerin.
6. Define emollient.

INHALATIONS

The term inhalation is used commonly by the layman to present preparations intended to be vaporized with the aid of heat, usually steam, and inhaled. Inhalations are solutions of medicaments administered by the nasal or oral respiratory route for local or systemic effect. Nebulizers are suitable for administration of inhalation solutions only if they give droplets sufficiently fine and uniform in size so that the mist reaches the bronchitis.

Another category of products also called inhalations or sometimes they are known as insufflations. They consist of finely powdered or liquid drugs that are carried into the respiratory passages by the use of special delivery systems. Solutions may be nebulized by use of inert gases. Nebulized solutions may be breathed directly from the nebulizer or nebulizer maybe attached to a plastic face mask or intermittent positive pressure breathing machines. Particle size is one of the important parameter for inhalation. The optimum particle size for penetration into the pulmonary cavity is 0.5 to 7.0 µm. Fine mists are generated by pressurized aerosols and hence posses basic advantages over the older nebulizers.

EXERCISE NO. 1

Object : To prepare Aqueous inhalation

Requirements : Eucalyptus oil, menthol, magnesium carbonate, purified water, etc.

Formula

Eucalyptus oil	10.0 ml
Menthol	3.0 g
Light magnesium carbonate	6.50 g
Purified water, sufficient to produce	100.0 ml

Procedure

1. Powder weighed amount of menthol in glass mortar and pestle.
2. Add eucalyptus oil and stir until the solid has dissolved.
3. Add light magnesium carbonate and mix well.

4. Add purified water to convert in a pourable cream.
5. Transfer to a tared container.
6. Rinse the mortar with purified water and add the rinsings to the bottle in small amounts.
7. Shake vigorously after each addition.

Category : Expectorant, bacteriostatic

Storage : Preserve in a well-closed container.

Uses : Expectorant, bacteriostatic

EXERCISE NO. 2

Object : To prepare Benzoin inhalation

Requirements : Benzoin, prepared storax, specially denatured spirit, etc.

Formula

Benzoin	10.0 g
Prepared storax	6.85 g
Specially denatured spirit, sufficient to produce	100.0 ml

Procedure

1. Benzoin and prepared storax should be macerated with 7.5 ml of specially denatured spirit for twenty four hours.
2. Filtere and add specially denatured spirit passed through filter to produce 100 ml.

Category : Anti-inflammatory

Storage : Preserve in a well-closed container.

Uses : As an anti-inflammation agent used for throat and bronchial inflammation

EXERCISE NO. 3

Object : To prepare Menthol inhalation

Requirements : Menthol, specially denatured spirit, etc.

Formula

Menthol	1.0 g
Specially denatured spirit, sufficient to produce	50.0 ml

Procedure

1. Mix menthol with specially denatured spirit in mortar and pestle.
2. Add sufficient specially denatured spirit through filter and produce 50 ml.

Category : Antipruritic

Storage : Preserve in a well-closed container.

Uses : As an antipruritic agents used for nasal congestion, headache and in neuralagia.

VIVA VOCE QUESTIONS

1. What are inhalations?
2. What are sprays?
3. What is the difference between inhalations and sprays.
4. What are insufflations?
5. What is the difference between insufflations and inhalations?
6. What is the use of menthol?
7. Give advantages of inhalations.
8. Why you prefer inhalations than application?
9. In which condition sprays are very important?

MOUTHWASHES

A mouthwash is an aqueous solution which is most often used for its deodorant, refreshing or antiseptic effect. It may contain alcohol, glycerin, synthetic sweeteners, surface active agent, flavoring and coloring agents. Mouthwashes generally contain following substances.

1. Antibacterial agents : Alkaline phenol, hydrogen peroxide, buffered sodium perborate, thymol glycerin.
2. Astringents : Zinc sulphate, zinc chloride, etc.

Examples

1. Phenol and alkaline mouthwashes
2. Potassium chlorate and phenol mouthwashes
3. Thymol glycerin mouthwashes
4. Hydrogen peroxide mouthwashes
5. Buffered sodium perborate mouthwashes

CONTAINER

Narrow mouthed screw capped colored fluted bottle.

LABELLING

Comply with the general requirements for labelling. In addition the label on the container states.

1. 'Not to be Swallowed in large amount'
2. 'Store in cool place'

EXERCISE NO. 1

Object : To prepare Compound sodium chloride mouthwash

Requirements : Sodium bicarbonate, sodium chloride, concentrated peppermint emulsion, double strength chloroform water, purified water, etc

Formula

Sodium bicarbonate	1.0 g
Sodium chloride	1.5 g
Concentrated peppermint emulsion	2.5 ml
Double strength chloroform water	50.0 ml
Purified water	100.0 ml

Procedure

1. Dissolve sodium bicarbonate and sodium chloride in purified water.
2. Add concentrated peppermint emulsion and mix.
3. Add double strength chloroform water.
4. Add sufficient purified water to produced 100 ml.

Category : antipruritic

Storage : Preserve in a well-closed continer.

Uses : antipruritic

VIVA VOCE QUESTIONS

1. What are mouthwashes?
2. What is difference between mouthwashes and gargles?
3. How to prepare mouthwashes?
4. What is the use of mouthwashes?
5. What is the purpose of mouthwashes?

EXERCISE NO. 2

Object : To prepare mouthwash

Requirements : Cetylpyridinium chloride, citric acid, sweetener (sodium saccharin), flavor oils (peppermint, eucalyptus, and clove oils), polyoxyethylene (20) sorbitan monostearate, ethanol, sorbitol solution and purified water, etc.

Theory : Cetylpyridinium chloride which is used as an antimicrobial agent and also contributes to the solubilizing power of the system. This preparation contain low concentaration of surfactant because the sorbitol contributes the solubilization efficacy of this mouthwash. The product illustrates the use of alcohol in the presence of a solubilizer to increase flower impact.

Formula

Cetylpyridinium chloride	0.10 g
Citric acid	0.10 g
Sweetener (sodium saccharin)	0.04 g
Flavor oils (peppermint, eucalyptus, and clove oils)	0.15 ml
Polyoxyethylene (20) sorbitan monostearate,	0.3 g
Ethanol	10.0 ml
Sorbitol solution	20.0 g
Purified water, sufficient to produce	100.0 ml

Procedure

1. Dissolve cetylpyridinium chloride, citric acid, and sodium saccharin in a sufficient amount of the water and add ethanol.
2. Mix polyoxyethylene (20) sorbitan monostearate and flavor oils.
3. Add slowly hydroalcoholic solution with stirring.
4. Add sorbitol and mix.
5. Add sufficient amount of purified water to produce 100 ml.

Category : Antimicrobial **Storage :** Preserve in a well-closed container.
Uses : Local anti-infective, and antiseptic

VIVA VOCE QUESTIONS

1. What are mouthwashes?
2. Why surfactant is used in mouthwashes?
3. How mouthwash is prepared?
4. What will happen if mouthwash is taken orally?

PAINTS

Paints are solutions or dispersions of one or more medicaments intended for application to the skin or in some cases mucous membrane. They may contain volatile solvent that evaporates quickly to leave a dry or resinous film of medicaments.
Throat paints are more viscous due to high content of glycerin. Paints are sticky and adheres to the affected site and prolong the action of the medicaments.

Examples

1. Brilliant green and crystal violet paints 2. Crystal violet paints 3. Coal tar paint 4. Compound mastric paint 5. Mandle's paint 6. Tannic acid glycerin paints

STORAGE

Paints should be kept in airtight containers.

LABELLING

Comply with the general requirements for labelling. In addition it should states 'For External Use Only'.

CONTAINER

Wide mouth screw capped colored bottles.

EXERCISE NO. 1

Object : To prepare Brilliant green and crystal violet paints

Requirements : Brillient green, crystal violet, ethanol (90%), purified water, etc.

Formula

Brillient green	0.5 g
Crystal violet	0.5 g
Ethanol (90%)	50.0 ml
Purified water, sufficient to produce	100.0 ml

Procedure

1 Dissolve the brilliant green and crystal violet in the ethanol (90%).
2 Add sufficient water to produce 100 ml.

Weight per ml : 0.930 to 0.950 g **Storage :** Preserve in a well-closed container.

Uses : Skin disease

EXERCISE NO. 2

Object : To prepare Crystal violet paints

Requirements : Crystal violet, purified water, and common glasswares.

Formula

Crystal violet	0.5 g
Purified water, sufficient to produce	100.0 ml

Procedure

1. Dissolve crystal violet in purified water.
2. Add sufficient purified water to produce 100 ml.

Storage : Crystal violet paint should be stored at a temperature not exceeding 25°C.

Uses : Skin disease

EXERCISE NO. 3

Object : To prepare Coal tar paint

Requirements : Coal tar, acetone, etc.

Formula

Coal Tar	10.0 g
Acetone, sufficient to produce	100.0 ml

Procedure

1. Disperse the coal tar in 70 ml of acetone and allow to stand for one hour.
2. Filter if necessary and dilute with acetone to 100 ml.

Category : Local irritant

Weight per ml : 0.810 to 0.820

Storage : Paint should be stored at a temperature not exceeding 25°C. The label on the container states that the preparation is flammable and that it should be kept away from naked flames.

Uses : A local irritant used in the treatment of chronic skin disease

EXERCISE NO. 4

Object : To prepare Compound mastric paint

Requirements : Mastric, castor oil, acetone, industrial methylated spirit, etc

Formula

Mastric	40.0 g
Castor oil	1.25 ml
Acetone	qs
Industrial methylated spirit, sufficient to produce	qs

Procedure

1. Dissolve the mastric in 45 ml of a mixture of equal volumes of acetone and industrial methylated spirit.
2. Add the castor oil and mix.
3. Add sufficient quantity of a mixture of equal volumes of acetone and industrial methylated spirit to produce 100 ml.
4. Filter and preserve in a suitable container.

Storage : Preserve in a well-closed container at a temperature not exceeding 25°C. The label on the container states that the preparation is flammable and that it should be kept away from naked flames.

Uses : Skin diseases

EXERCISE NO. 5

Object : To prepare Iodine paint

Synonyms : Mandl's paint

Requirements : Potassium iodide, Iodine, ethanol (90%), peppermint oil, purified water, glycerin, etc

Theory : Paints are liquid preparations intended for application to the mucous membrane or skin. They usually contain substances possessing analgesic, antiseptic, caustic or astringent properties. In this preparation Iodine acts as antiseptic, potassium iodide dissolves the iodine. Peppermint oil acts as flavoring agent and produce cooling effect. Alcohol is used as a solubilising agent of the peppermint oil. This preparation contains iodine, hence should be prepared in glass apparatus. It should not be prepared in mortar and pestle because it contains the pores and iodine enters in these pores. It is difficult to washout the iodine and this entangled iodine can change the color of the other preparation. The glycerin is used as vehicle in paints. Glycerin is very important vehicle and used in the paints due to the following reasons.

1. It is an excellent solvent for those substance which are not soluble in aqueous medium such as boric acid, tannic acid, phenol, borax, etc.
2. Due to viscous nature, it adhere stir long period to the skin or mucous membrane.
3. It is hygroscopic in nature. Due to this reason, the area on which the preparation is applied remains in moist condition.
4. It acts as a preservative due to higher osmotic pressure. It kills the microorganisms by withdrawing water from them by producing dehydration.
5. Glycerin is sweet in taste and acts as a demulcent.
6. It acts as local soothing effect on the unbroken sin or mucous membrane.

Formula

Potassium iodide	2.5 g
Iodine	1.25 g
Ethanol (90%)	4.0 ml
Peppermint oil	0.4 ml
Purified water	2.5 ml
Glycerin, sufficient to produce	100.0 ml

Procedure

1. Dissolve potassium iodide and iodine in purified water in glass mortar and pestle.
2. Add a small portion of glycerin.
3. Add peppermint oil dissolve in ethanol and mix.
4. Add sufficient glycerin to produce 100 ml.

Category : Antiseptic

Storage : Store in airtight container in cool place because it contains volatile substances peppermint oil and ethanol. It is also protected from atmospheric moisture because glycerin is hygroscopic in nature.

Label :
'Store in a Cool Place' because iodine is volatile
'Shake the bottle before use' because some of oil separates on storage.
'Not to be Swallowed in Large Amounts'

Uses : Antiseptic throat paint.

VIVA VOCE QUESTIONS

For Exercise No. 27 to 31

1. What are paints?
2. How to prepare paints?
3. What is the use of paints?
4. What is Mandl's paints?
5. Which types of skin diseases are treated with paints?
6. What is the use of iodine?
7. What is the use of glycerin in the preparation of paints?
8. Give storage condition of paints.
9. Why viscous paints are prepared?
10. How to apply paints?
11. Paints are suitable for all types of skin disease or not?
12. Why potassium iodide is essential for the preparation of Iodine paints?

PASTES

Pastes are semi-solid preparations usually containing one or more medicaments in suitable bases. They usually contain a high proportion of solids as compare to ointments. They are usually employed to apply medicaments to small localized area. Pastes should be supplied in collapsible tubes of metal or plastic or in glass or other suitable containers which prevents the evaporation of the volatile additives.

EXERCISE NO. 1

Object : To prepare Compound aluminium paste

Synonyms : Baltimore Paste

Requirements : Aluminium powder, zinc oxide, liquid paraffin, etc.

Formula

Aluminium powder	20.0 g
Zinc oxide	40.0 g
Liquid paraffin	40.0 g

Procedure :

1. Mix the aluminium powder and zinc oxide with liquid paraffin until smooth.

Category : Mild astringent, protective and antiseptic
Storage : Preserve in a aliminium collapsible tube or wide mouth bottle
Uses : Mild astringent, protective and antiseptic

EXERCISE NO. 2

Object : To prepare Resorcinol and sulphur paste

Requirements : Resorcinol, precipitated sulphur, zinc oxide, emulsifying ointment, etc

Formula :

Resorcinol, finely sifted	5.0 g
Precipitated sulphur	5.0 g
Zinc oxide, finely sifted	40.0 g
Emulsifying ointment	50.0 g

Procedure

1. Prepare emulsifying ointment.
2. Triturate the resorcinol, precipitated sulphur and zinc oxide with a portion of the emulsifying ointment until smooth.
3. Gradually incorporate the remaining quantity of the emulsifying ointment.

Category : Anti-fungal, antibacterial, protective

Storage : Preserve in a suitable container. The label on the container states that the paste should be protected from light.

Uses : Anti-fungal, antibacterial, protective

EXERCISE NO. 3

Object : To prepare Compound zinc paste

Requirements : Zinc oxide, starch, white soft paraffin, etc.

Formula :

Zinc oxide, finely sifted	25.0 g
Starch, finely sifted	25.0 g
White soft paraffin	50.0 g

Procedure

1. Melt white soft paraffin in a glass container.
2. Incorporate zinc oxide and starch and stir until cold.

Category : Protective, astringent and antiseptic

Storage : Preserve in a well-closed container.

Uses : Protective, astringent and antiseptic

EXERCISE NO. 4

Object : To prepare Coal tar paste

Requirements : Strong coal tar solution, compound zinc paste, etc

Formula

Strong coal tar solution	7.5 g
Compound zinc paste	9.25 g

Procedure

1. Prepare strong coal tar solution and compound zinc paste.
2. Triturate strong coal tar solution with a portion of the compound zinc paste until smooth.
3. Gradually add the remaining amount of the compound zinc paste.

Category : Protective
Storage : Preserve in a well-closed container.
Uses : Protective

EXERCISE NO. 5

Object : To prepare Zinc and salicylic acid paste

Synonyms : Lassar's Paste

Requirements : Zinc oxide, salicylic acid, starch, white soft paraffin, etc.

Formula :

Zinc oxide, finely sifted	24.0 g
Salicylic acid, finely sifted	2.0 g
Starch, finely sifted	24.0 g
White soft paraffin	50.0 g

Procedure

1. Melt white soft paraffin in a glass container.
2. Incorporate the zinc oxide salicylic acid, and starch stir until cold.

Category : Protective

Storage : Preserve in a well-closed container.

Uses : Protective

EXERCISE NO.

Object : To prepare Zinc and coal tar paste

Requirements : Emulsifying wax, zinc oxide, coal tar, starch, yellow soft paraffin, etc.

Formula :

Emulsifying wax	5.0 g
Zinc oxide, finely sifted	6.0 g
Coal tar	6.0 g
Starch	38.0 g
Yellow soft paraffin	45.0 g

Procedure

1. Melt the emulsifying wax at 70°C.
2. Add coal tar and 22.5 g of yellow soft paraffin and stir at 70° until completely melted.
3. Add the remainder of the yellow soft paraffin and cool at 30°.
4. Add zinc oxide and the starch with constant stirring and stir until cold.

Category : Protective, astringent and antiseptic

Storage : Preserve in a well-closed container.

Uses : Protective, astringent and antiseptic

EXERCISE NO. 7

Object : To prepare Titanium dioxide paste

Requirements : Titanium dioxide, chlorocresol, red ferric oxide, light kaolin, zinc oxide. glycerol. purified water, etc.

Formula

Titanium dioxide	20.0 g
Chlorocresol	0.10 g
Red ferric oxide	2.0 g
Light kaolin (sterilized)	10.0 g
Zinc oxide	25.0 g
Glycerol	15.0 g
Purified water, sufficient to produce	100.0 g

Procedure

1. Powder titanium dioxide, red ferric oxide, light kaolin, and zinc oxide.
2. Dissolve the chlorocresol in glycerol.
3. Add purified water gradually and triturate the solution.
4. Mix the powders and triturate until a smooth paste.

Category : Anti-bacterial. protective

Storage : Preserve in a well-closed container.

Uses : Anti-bacterial, protective

VIVA VOCE QUESTIONS

1. What are pastes?
2. Why paste is prepared?
3. What is ointment?
4. Define and differentiate cream. paste, gels, jellies. and paste.
5. What is ferric oxide red?
6. What are the applications of pastes?
7. Which types of materials are suitable for paste?
8. Which types of additives are essential forr pasttes?D

POULTICES

Poultices are viscous or paste-like preparations intended for external use only. They are applied to the skin. They may stimulate body surface or alleviate an inflamed area by supplying medicating substances in the presence of heat and moisture. Poultices must retain heat for a considerable time because they are intended to supply warmth to inflamed parts of body. They may contain following substances.

1. Absorbent : e.g. kaolin
2. Hygroscopic : e.g. glycerin
3. Anti-rheumatic : e.g. methyl salicylate
4. Bactericide : e.g. thymol, boric acid
5. Flavoring agent : Peppermint oil
6. Coloring agents

CONTAINER

Wide mouth screw capped colored bottle.

EXERCISE NO. 1

Object : To prepare Kaolin poultice

Requirements : Heavy kaolin, methyl salicylate, thymol, peppermint oil, glycerol, etc.

Formula

Heavy kaolin, finely sifted	52.7 g
Boric acid, finely sifted	4.5 g
Methyl salicylate	0.2 ml
Thymol	50.0 mg
Peppermint oil	0.05 ml
Glycerol	42.5 g

Procedure

1. Mix the heavy kaolin and the boric acid with the glycerol.
2. Heat the mixture at 120°C for one hours, stirring occasionally and allow to cool.(A)
3. Dissolve thymol and methylsalicylate in separate container.(B)
4. Add (B) in (A).
5. Add the peppermint oil and mix thoroughly.

Category : Germicide

Storage : Kaolin poultice should be supplied in suitable containers which reduce absorption, diffusion or evaporation of additives.

Uses : Local anti-infective

VIVA VOCE QUESTIONS

1. What are poultices?
2. What is difference in poultices and paint?
3. How to prepare poultices?
4. What is the use of poultices?
5. Give application of kaolin poulitices.

SPRAYS

Sprays are solutions of medicaments of various drugs in aqueous vehicles and are applied to the mucous membrane of the nose and the throat by means of an atomizer or nebulizer. Spray devices should produce relatively coarse droplets if the therapeutic action of the medicaments is to be restricted to the upper respiratory tract. Very fine droplets penetrate into the respiratory tract. Initially the spray are prepared by dissolving various medicaments in light liquid petrolatum. It may damage the normal ciliary activity of the nasal mucosa and, if drops of oil enter the trachea it can cause lipoid pneumonia. On the basis of this problem, aqueous sprays which are isotonic with nasal secretions and of approximately the same pH are preferred. Sprays contain following substances such as

1. Antibiotics
2. Antihistamines
3. Vasoconstracters
4. Alcohol
5. Solubilizers
6. Wetting agents

EXERCISE NO. 1

Object : To prepare Isoprenaline compound spray

Requirements : Isoprenaline, papaverine hydrochloride, atropine methonitrate, propylene glycol, sodium metabisulphite, purified water, etc.

Formula

Isoprenaline	0.250 g
Papaverine hydrochloride	0.625 g
Atropine methonitrate	50.0 mg
Propylene glycol	1.25 ml
Sodium metabisulphite	25.0 mg
Purified water, sufficient to produce	25.0 ml

Procedure

1. Mix isoprenaline, atropine methonitrate and papaverine hydrochloride and dissolve in purified water.
2. Dissolve sodium metabisulphite and add in the solution.
3. Add other additives and add sufficient purified water to produce 25 ml.

Category : Parasympatholytic

Storage : Preserve in a fluted narrow-mouthed, light-resistant, screw-capped bottle.

Uses : Parasympatholytic

EXERCISE NO. 2

Object : To prepare Compound adrenaline and atropine spray

Requirements : Adrenaline acid tartrate, atropine methonitrate, papaverine hydrochloride, sodium metabisulphite, chlorbutanol, propylene glycol, purified water, etc.

Formula

Adrenaline acid tartrate	0.80 g
Atropine methonitrate	0.10 g
Papaverine hydrochloride	0.80 g
Sodium metabisulphite	0.10 g
Chlorbutanol	0.50 g
Propylene glycol	5.00 ml
Purified water, sufficient to produce	100.0 ml

Procedure

1. Boiled purified water for about 10 minutes, to remove oxygen and destroy vegetative bacteria and mould spores. Cool the water and use for the preparation of sprays.
2. Prepare solution of adrenaline acid tartrate, atropine methonitrate and sodium metabisulphite in separate containers respectively using small proportion of purified water.
3. Mix three solution and add papaverine hydrochloride in the mixture.
4. Dissolve weighed amount of chlorbutanol in propylene glycol and mix in the mixture.
5. Filter, if necessary and add sufficient purified water to adjust the volume 100 ml through the filter.

Category : Parasympatholytic

Storage : Preserve in a fluted narrow-mouthed, light-resistant, screw-capped bottle because this spray is susceptible to oxidation. The container should be well filled.

Label : The label should state --

1. The preparation should be stored in a cool place.
2. A darkened solution must not be used.
3. The preparation should be discarded after not more than one month.

Uses : Parasympatholytic

VIVA VOCE QUESTIONS

1. What are sprays?
2. What is the use of atropine salt?
3. Why will you prepare sprays?
4. How sprays are used?
5. What is the pharmaceutical application of adrenaline salt?
6. Why sprays are stored in cool place?

APPENDICES

A Interconversions

Metric- Imperial equivalent

4 Milliliter	= 1.0 fluid drachm	1 Litre	= 35.20 fluid ounce
1 Gram	=15.43 grain	1 Milliliters	= 16.90 minim
1 Kilogram	=2.205 pound	1 Gallon (gal)	= 4.549 litre
1 Ton	= 1016.0 kilogram	1 Gallon (US)	= 3.785 litre
1 Pound	= 453.59 gram	1 Pint	= 568.26 milliliters
1 Ounce (avoirdupois)	=28.35 gram	1 Fluid ounce	= 28.41 milliliters
1 Ounce (apothecary)	= 31.10 gram	1 Fluid drachma	= 3.552 milliliters
1 Grain	= 0.065 gram	1 Minim	= 0.059 milliliters

Common Household (Metric-Imperial Equivalents)

1 drop	= 1 minim	= 0.04 millilitre
1 teaspoonful	= 1 fluid dracham	= 4 millilitre
1 Desertspoonful	= 2 fluid dracham	= 8 millilitre
1 Tablespoonful	= 4 fluid dracham	= 15 millilitre
1 Wineglassful	= 2 fluid ounce	= 60 millilitre
1 Teacupful	= 4 fluid ounce	= 120 millilitre
1 Tumblerful	= 8 fluid ounce	= 240 millilitre

B. List of common glassware & other items

S.N.	Name	S.N.	Name
1.	Ampoules	20.	Round bottom flask
2.	Aluminium tube (Collapsible tube)	21.	Flat bottom flask
3	Beakers	22.	R.D.bottle
4.	Burettes (punch clip, glass stopper)	23.	Slides
5.	Burette clip (steel, teflon, iron)	24.	Separating funnel
6	Buchner funnel (porcelin)	25.	Spatula
7.	Conical flask	26.	Test tubes
8.	Perculators (steel, glass)	27.	Test tube stands
9.	China dishes	28.	Thermometer
10.	Clamps (steel, teflon)	29.	Test tube holders
11.	Dispensing bottle	30.	Tripod stand
12.	Desiccator (simple or vacuum)	31.	Vials
13.	Funnel (glass & plastic)	32.	Volumetric flask
14.	Filtration flask Glass tubes	33.	Weighing bottles
15.	Iodine flasks	34.	Water bath (constant level)
16.	Pestle mortar (porcelin, glass)	35.	Wire gauge
17.	Pipettes graduated	36.	Watch glasses
18.	Pipette with bulb	37.	Wash bottles
19.	Petri dishes	38.	Measuring cylinder

C. List common equipments

S.No.	Name of Equipment	S.No.	Name of Equipment
1.	Air curtain	36.	Hot plates
2.	Air shower	37.	Humidifier
3.	Aseptic cabinet	38.	Humidity cabinet
4.	Antibiotic zone reader	39.	Hot air oven
5.	Autoclave	40.	Incubator
6.	Air conditioners	41.	IR moisture balance
7.	Incubator shaker	42.	Knife sharpener
8.	Water bath	43.	Laminar flow bench
9.	Biological safety cabinet	44.	Microscopes
10.	Blood bank refrigerator	45.	Metabolic shaker
11.	B.O.D. incubator	46.	Micro tissue homogenizer
12.	Bacteriological filtration assembly	47.	Ointment filling machine
13.	Carbon dioxide incubator	48.	Ointment tube crimping machine
14.	Capsule filling machine	49.	Orbital shaking incubator
15.	Colony counter	50	Pipette dryer
16.	Coating pan	51	Pipette washer
17.	Centrifuge (1000 –5000 rpm)	52.	Rotary flask shaker
18.	Conductivity meter	53.	Seitz filter
19.	De-humidifier	54.	Tablet machine
20.	Desiccator cabinet	55.	Test tube rotator
21.	Distillation apparatus	56.	Turbidity meter digital
22.	Disintegration apparatus	57.	UV-Analysis cabinet
23.	Dissolution apparatus	58.	Ultra Violet lamp
24.	Environmental chamber	59.	UV-chamber
25.	Environmental growth cabinets	60.	Vacuum oven
26.	Fuming- chamber	61.	Vacuum cleaner
27.	Friabilator	62.	Vortex shaker
28.	Filter sterilization equipment	63.	Refrigerator
29.	Fluorescent lamps	64.	Sterility Test apparatus
30.	First-aid box	65.	Wrist action shakers
31.	Glassware Dryer	66.	Vortex apparatus
32.	Granulator	67.	Gyratory shaker
33.	Hardness testers (Monsanto or Pfizer tester)	68.	Double distilled water equipment
34.	Heating mental	69.	Trays and baskets
35.	Homoginizer	70.	Plastic carboys